Advances in Intelligent Systems: Reviews

Book Series, Volume 1

S.Yurish
Editor

Advances in Intelligent Systems: Reviews

Book Series, Volume 1

International Frequency Sensor Association Publishing

S. Yurish, *Editor*
Advances in Intelligent Systems: Reviews, Book Series, Vol. 1

Published by IFSA Publishing, S. L., 2017
E-mail (for print book orders and customer service enquires):
ifsa.books@sensorsportal.com

Visit our Home Page on http://www.sensorsportal.com

Neither the authors nor International Frequency Sensor Association Publishing accept any responsibility or liability for loss or damage occasioned to any person or property through using the material, instructions, methods or ideas contained herein, or acting or refraining from acting as a result of such use.

e-ISBN: 978-84-697-8923-0
ISBN: 978-84-697-8924-7
BN-20171226-XX
BIC: UYD

Contents

7

Preface

It is my great pleasure to present the first volume of the Advances in Intelligent Systems: Reviews book Series started by the IFSA Publishing in 2017.

'Advances in Intelligent Systems: Reviews' Vol. 1 Book Series is covering some design and architectural aspects related to intelligent systems and software. It ranges from the microarchitecture level via the system software level up to the application-specific architecture level.

The book volume contains ten chapters written by 25 contributors from academia and industry from 8 countries: Colombia, Denmark, France, Germany, Italy, Japan, Romania and USA.

Chapter 1 reports new results on parallel optimization for distributed intelligent systems. A self-optimization approach is studied that does not only consider pure load-balancing but also takes into account trust to improve the assignment of important services to trustworthy nodes. Authors use different optimization strategies to determine whether a service should be transferred to another node or not. Based on extensive simulations, it was shown that the proposed approach is able to balance the workload between nodes nearly optimal. Moreover, it improves significantly the availability of important services. The new results can be used as guidelines for influencing the self-optimization strategies to adapt to the specific environment of a concrete use case.

Chapter 2 describes a task mapping proposal, based on the Population-Based Incremental Learning (PBIL) algorithm. Obtained results show that PBIL may be used as a feasible optimization approach, which exhibits a better tradeoff between quality of the mapping solutions, and convergence time, when compared to some other reported optimization alternatives. Results also highlight the use of the adaptive feature of the PBIL algorithm, which performs better by using linear or sigmoidal learning rules.

Chapter 3 presents a new methodology for development of simulation applications, which elaborates the concept of dynamic simulation and analyses the major requirements towards creating a software platform implementing this concept on a wide range of the hardware devices. The combination of simulation services into a rich-functional simulation platform, as enabled by the software implementation technologies that were discussed in the chapter, offers a new perspective of the simulation technology evolution.

Chapter 4 describes the adaptive subdivision method for improvement of the calibration uncertainty, which allows the cylindrical weights with a lifting knob, having nominal values of (500...100) g, to be calibrated using an automatic comparator (which is not equipped with weight support plates). The method can be used for class E1 weights, where the highest accuracy is required. In this case, the resulting calibration uncertainty for the unknown weights is better than that usually obtained for E1 masses, being at the level of reference standards.

Chapter 5 presents solution focuses on implementing risk and hazard control paradigm. The approach proposed in this chapter aims to integrate the emerging technology: Risk and Hazard Control as control strategies for uncertainty management. Control Strategy is a planned set of controls, checks and sequences, derived from current product and process, having the target to assure process performance and product quality.

Chapter 6 describes a version of safe recursion, together with constructive diagonalization; by means of these two operators, it have been able to define a hierarchy of classes of programs. It was defined a hierarchy of programs with simultaneous time and space complexity bound.

Chapter 7 demonstrates how games can be regarded as actors and as organizers of actors and actions based on Actor Network Theory. The chapter is the result of a research project where authors studied players of different ages playing computer games, board games, and digital play equipment. The study is focused on studying games as a genre rather than just digital games, and our main example here is a board game.

Chapter 8 focus on composite armor structures consisting of several layers of ultra-high molecular weight polyethylene (UHMW-PE), a promising ballistic armor material due to its high specific strength and stiffness. The goal is to evaluate the ballistic efficiency of UHMW-PE composite with numerical simulations, promoting an effective development process.

Chapter 9 extends previous result by systematically reassessing the motion gains for the three lateral motion components for several levels of acceleration. A slalom task was chosen (with the level of lateral acceleration modified by changing the distance between posts) so that cornering behavior and self-motion perception could be assessed for various settings of the three parameters. Three motion components generally used to simulate lateral acceleration should be set individually and that use of the same motion gain for all three is not the best solution for improving the realism of the simulator.

Chapter 10 describes a step climbing strategy and a theoretical analysis method of step climbing for a typical wheelchair and indicated the difficulty of step climbing. Author has shown the cooperative step climbing strategy of the wheelchair and the robot, the robot and wheelchair system, and the theoretical analysis.

We hope that readers enjoy this book and that can be a valuable tool for those who involved in research and development of various intelligent systems.

Sergey Y. Yurish

Editor
IFSA Publishing   Barcelona, Spain

Contributors

Jose Aedo
Electronics and Communication Engineering Department
Universidad de Antioquia, Colombia

Nader Bagherzadeh
The Henry Samueli School of Engineering University of California,
Irvine, USA

Freddy Bolanos
Electrical Engineering and Automatics Department Universidad
Nacional de Colombia, Colombia

Christophe Bourdin
Aix Marseille Univ., CNRS, ISM, Marseille, France

Alexey Cheptsov
High Performance Computing, Center Stuttgart, Germany

Emanuele Covino
Dipartimento di Informatica, Università di Bari, Italy

Emmanuelle Diaz
PSA Groupe, Centre Technique de Vélizy, Route de Gizy 78943
Vélizy-Villacoublay Cedex, France

Gheorghe Florea
SIS, SA, Bucharest, Romania

Vincent Honnet
PSA Groupe, Centre Technique de Vélizy, Route de Gizy 78943
Vélizy-Villacoublay Cedex France

Hidetoshi Ikeda
Department of Mechanical Engineering, National Institute of
Technology, Toyama College, Japan

Alexandra Ionescu
SIS, SA, Bucharest, Romania

Carsten Jessen
Centre for Teaching Development and Digital Media, Aarhus University, Denmark

Jari Due Jessen
Center for Playware, Technical University of Denmark, Denmark

Stéphane Masfrand
PSA Groupe, Centre Technique de Vélizy, Route de Gizy 78943 Vélizy-Villacoublay Cedex France

Nizar Msadek
Department of Computer Science, University of Augsburg, Germany

Giovanni Pani
Dipartimento di Informatica, Università di Bari, Italy

Arash Ramezani
Institute of Automation Technology, University of the Federal Armed Forces Hamburg, Holstenhofweg 85, 22043 Hamburg, Germany

Fredy Rivera
Computer Science Department, Universidad de Antioquia, Colombia

Hendrik Rothe
Institute of Automation Technology, University of the Federal Armed Forces Hamburg, Holstenhofweg 85, 22043 Hamburg, Germany

Vincent Roussarie
PSA Groupe, Centre Technique de Vélizy, Route de Gizy 78943 Vélizy-Villacoublay Cedex France

Florian Savona
Aix Marseille Univ, CNRS, ISM, Marseille, France

Anca Stratulat

PSA Groupe, Centre Technique de Vélizy, Route de Gizy 78943 Vélizy-Villacoublay Cedex, France

Theo Ungerer

Department of Computer Science, University of Augsburg, Germany

Adriana Vâlcu

Regional Directorate of Legal Metrology, Bucharest, Romania (Formerly: National Institute of Metrology, Bucharest, Romania)

Philippe Vars

PSA Groupe, Centre Technique de Vélizy, Route de Gizy 78943 Vélizy-Villacoublay Cedex France

1.

Parallel Optimization for Intelligent Systems: Principles and New Results

Nizar Msadek and Theo Ungerer

1.1. Introduction

Intelligent distributed systems are rapidly getting more and more complex. Therefore, it is essential that such systems will be able to adapt autonomously to changes in their environment. They should be characterized by so-called self-* properties such as self-configuration [1-3], self-optimization [4-6] and self-healing [7, 8]. The autonomous optimization of nodes at runtime in open distributed environments is a crucial part for developing self-optimizing systems. In this chapter, a trust-aware self-optimization algorithm for self-* systems is presented. It does not only consider pure load-balancing but also takes into account trust to improve the assignment of important services to trustworthy nodes. The proposed self-optimization approach makes use of different optimization strategies based on trust to determine at runtime whether a service should be transferred to another node or not. The trust definition [9] adopted for this work is the definition provided by the research unit OC-Trust of the German Research Foundation (DFG) by regarding different facets of trust, as, for example, safety, reliability, credibility and usability. The focus here lies on the reliability aspect. Furthermore, it is assumed that a node can not realistically assess its own trust value because it trusts itself fully. Therefore, the calculation of the trust value in this work must be done with the previously introduced trust metrics presented in [10]. With trust information, nodes of a system have a reference about which nodes to cooperate with, and this is important for self-optimizing systems. The chapter offers as contribution the following aspects:

Nizar Msadek
Department of Computer Science, University of Augsburg, Germany

1) A decentralized self-optimization algorithm for load balancing taking into account trust — respectively reliability — to increase the robustness of important services in open distributed environments (see Sections 1.3 and 1.4),

2) A formal description of the optimization strategies to determine at runtime whether a service should be transferred to another node or not (see Section 1.5), and

3) A set of extensions for the basic algorithm to further improve its performance time in case of multiple simultaneous requests (see Section 1.6).

All aspects are evaluated and discussed with respect to a toolkit based on the TEM [11], a trustenabling middleware for building real-world distributed Organic Computing systems. Section 1.7 provides evaluation results of the proposed self-optimization algorithm and demonstrate the benefits of the proposed extensions. Finally, the chapter is closed with a conclusion and future work in Section 1.8.

1.2. Related Work

A lot of papers have been published to deal with the assignment problem of services on nodes, either to achieve a static or dynamic load balancing [12-17]. In most existing algorithms, the consideration of the trustworthiness of nodes has been neglected so far. For instance, the work of Rao et al. [18] proposes several methods for solving the load balancing problem in distributed systems. One of these methods, called one-to-one, is similar to our approach: two nodes are picked at random. Then, a virtual server transfer is initiated if one of the nodes is heavy and the other is light. Their method, however, does not consider how the availability of important services may be improved, and does not distinguish between trustworthy and untrustworthy nodes. Bittencourt et al. [19] presented an approach to schedule processes composed of dependent services onto a grid. This approach is implemented in the Xavantes grid middleware and arranges the services in groups. It has the drawback of a central service distribution instance and therefore a single point of failure can occur. In [20], two different self-optimization algorithms for LTE networks are presented. One of these algorithms, called Load Balancing in Downlink LTE networks, is similar to our approach. The authors try to shift the virtual load of overloaded cells to less loaded adjacent cells by changing the virtual cell borders. The virtual load is modeled as the sum of resources needed to achieve a certain QoS for all active user equipment. Matrix [21] is another approach to combine load optimization with data-aware scheduling. The authors propose to apply adaptive work stealing techniques to achieve load balancing in distributed many-tasks computing environment. Tasks are organized in queues based on their size and locations. Then, a ZHT is used to submit tasks to idle schedulers and to monitor the execution progress of tasks in a scalable way. Whenever a scheduler has no more tasks, it communicates with other heavy-loaded schedulers to receive new tasks. Their approach does not take the priority of different service classes into account. In [22], the authors presented a receiver-initiated optimization algorithm that automatically balances the workload of nodes in distributed computing environments. It is implemented in the $OC\mu$ middleware. In their algorithm, services can be relocated or transferred to other nodes to balance the resource consumption among nodes. Moreover, it takes the trust constraints of nodes into account to transfer important services only to trustworthy nodes. However, it is based on the unrealistic assumption that all nodes have the same resource capacity. Contrary to this work, our approach is able to work with heterogeneous capacities. More precisely, we are interested in a dynamic receiver-initiated [23] self-optimization algorithm (i.e., since services are assumed not to be stolen from other nodes) that has neither a central control nor complete knowledge about the system. The algorithm must not only consider pure load-balancing but also takes into account trust to improve the assignment of important services to trustworthy nodes. And all this at runtime.

1.3. Basic Idea of the Self-Optimization Algorithm

A distributed system consisting of a set of n nodes $N = \{n_1, n_2.., n_n\}$ is considered, where each node can interact with each other through a set of application messages. They can optimize at runtime the assignment of services in the network by transferring their own services to other nodes. Suppose that node j at a certain point during runtime sends an application message to another node i. It appends onto the outgoing message (a) its trust in node i (b) its current workload and (c) some information (i.e., importance level and consumption) about services, which are running on it. Based on this information node i decides which of the following optimization strategies should be performed:

1.3.1. No Optimization

Description: The workload between nodes is well balanced and their trust values are similar enough.

Discussion: This is the simplest case that can happen between nodes. Both of them are well optimized in terms of trust and workload.

Solution: Nothing will happen

1.3.2. Load Optimization

Description: Trust of nodes is similar enough but their workload is unbalanced.

Discussion: This strategy aims to find a pure load balancing between nodes since their trust is similar enough.

Solution: Services are transferred in order to balance the workload between the nodes. Then, two cases are distinguished: (a) either the workload of i is higher or (b) the workload of j is higher. In the case of (a), node i balances the workload of the nodes by transferring a subset of its services to j. Otherwise, node i sends an alert message to j together with all information which are necessary for the optimization. Case (a) will be then triggered on side of j.

1.3.3. Trust Optimization

Description: The workload between nodes is well balanced but their trust values differ significantly. In this case important services might run on untrustworthy nodes and are prone to fail.

Discussion: This strategy aims to use particularly trustworthy nodes for important services. Therefore, important services have to be relocated to more trustworthy nodes and unimportant services to less trustworthy nodes. Furthermore, the overall workload resources between nodes should still be well-balanced.

Solution: By this strategy, we distinguish between two cases: (a) either i is more trustworthy than j or (b) j is more trustworthy than i. If (a), then i swaps its unimportant services for important services of j. In the case of (b), node i swaps its important for unimportant services of j. Note that the load consumption between important and unimportant services should be similar to keep the load-balancing property in both nodes satisfied.

1.3.4. Trust and Load Optimization

Description: Trust of nodes differs significantly and their workload is unbalanced.

Discussion: This strategy aims at workload balancing with additional consideration of the services' priority, i.e. to avoid hosting important services on untrustworthy nodes.

Solution: Four cases are distinguished: (a) Either the workload of i is higher and i is more trustworthy than j, (b) The workload of i is higher but j is more trustworthy, (c) The workload of j is higher but it is less trustworthy than i, or finally (d), The workload of j is higher and it is also more trustworthy than j. In the case of (a), node i balances the workload of load by transferring only unimportant services to j. If there are no unimportant services available, then no optimization is done. The rationale for this step is that there is a trade-off between trust and workload. Improving one of these criteria will typically deteriorate the other. In the case of (b), node i balances the workload by transferring only important services to j. Just as the case of (b), no optimization is done, if there are no available unimportant services. In other cases (i.e., c and d), node i sends an alert optimization message to j to piggy-back information necessary for self-optimization. Depending on the situation, case (a) or (b) will be then triggered on side of j.

1.4. Metrics and Notions

Since it is very complex to address the self-optimization problem in its full generality, we make some simplifying assumptions. Firstly, we assume that the load of a service is stable (or can otherwise be predicted) over the time interval it takes for the self-optimization algorithm to operate. Secondly, we assume there is only one bottleneck resource we are trying to optimize for. Let w_i denote the workload of a node i, where w_i represents the sum of the resource consumptions of all services running on node i (see Formula 1.1).

$$w_i = \sum_{s \in S_i} c_s, \text{ with } 0 \le w_i \le C_i^{max} \tag{1.1}$$

It is to note that c_s means the resource consumption of a service s. The maximum resource capacity of a node i is denoted by C_i^{max} and its set of services by S_i. Moreover, we divide services S_i into two sets based on their importance levels:

- S_i^{imp}: Set of important services (running on node i), which are necessary for the functionality of the entire system.

- S_i^{unimp}: Set of unimportant services (running on node i), which have only a low negative effect on the entire system if they fail.

Then, considering only the context of pure load optimization, our goal is to balance the workload between nodes. Let us assume two nodes, i and j: node i is underloaded. However, node j is overloaded and its task is to balance the workload by service transfers to i. Thus, as you can see Fig. 1.1: Simple load optimization method in Fig. 1.1, j transfers its services whose cumulative resource consumption is close enough to $\frac{|w_j - w_i|}{2}$ (optimal balancing). Although this simple idea seems to make a lot of sense, its drawback arises when the resource capacities of nodes are significantly different (see Fig. 1.2).

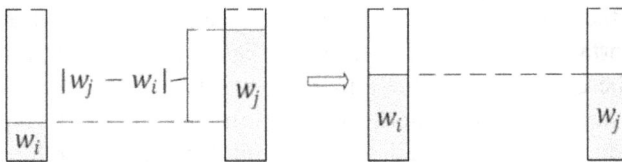

Fig. 1.1. Simple load optimization method.

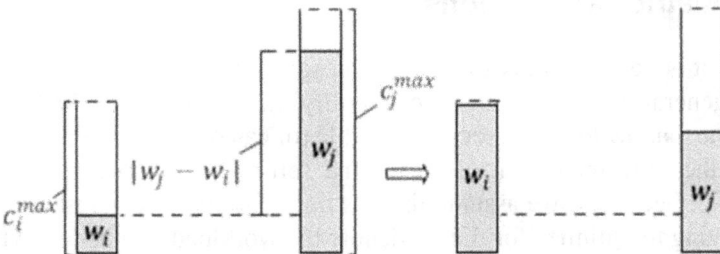

Fig. 1.2. Nodes still unbalanced due to their different resource capacities.

$$Q_i = \frac{w_i + w_j}{c_i^{max} + c_j^{max}} c_i^{max} \tag{1.2}$$

Therefore, we introduce a new optimal theoretical workload O_i, which should serve as a target reference point for every node. The node which surpasses this reference point ($w_i > O_i + \delta_{tol}$) is considered to be overloaded, otherwise it is underloaded ($w_i < O_i - \delta_{tol}$) or balanced ($|O_i - w_i| \le \delta_{tol}$), where a δ_{tol} is a tolerable threshold and represents the quality to reach the perfect workload. The optimal theoretical workload of a node i is calculated using Formula 1.2. Since w_i is normalized in a different capacity than w_j, we must first divide the sum of workload $w_i + w_j$ by the sum of capacity $c^{max}_i + c^{max}_j$ to obtain the optimal theoretical workload per one unit capacity, which will be then multiplied by c^{max}_i. Furthermore, each node has an individual trust value calculated based on the previously introduced trust metrics presented in [10]. Recall, the trust value $t_i(j)$ represents the subjective trust of node i in node j and will always range between 0 and 1. The value of 0 means that i does not trust j at all while a value of 1 stands for complete trust. Two nodes i and j are considered to have a similar trust behavior if $|t_i(j) - t_j(i)| \le \gamma_{tol}$, where γ_{tol} is a tolerable threshold and reflects the quality to achieve a good trust similarity between nodes.

1.5. The Algorithm in Detail

The algorithm proposed in this section represents a best-effort approach to improve the assignment of services on nodes so as to satisfy both workload and trust constraints. It is used to solve this problem in a distributed manner. We assume that nodes of the network do not know the workload of others until they receive a message from a node with information about that. The workload of nodes also might change over time. We further assume that a node can not assess its own trust value, but is rated by other nodes. Therefore, its trust value must be calculated from the neighbor nodes of the network (see [10] for more details). Note that the trust of nodes might also change over time. Again we are considering two nodes i and j, where j sends an application message m_j to i, on which it piggybacks the following additional information:

- S_j^{unimp}: Set of less important services running on node j
- S_j^{imp} : Set of important services running on j
- $t_j(i)$: Current trust value of j in i
- w_j: Current workload value of j

- $c^{max}{}_j$: Maximum resource capacity of j

Based on this information node i decides which optimization strategy should be performed. In the following we consider all possible decisions a node i has to make:

1.5.1. No Optimization

Formal description: $|t_i(j)-t_j(i)| \leq \gamma_{tol}$ and $|O_i - w_i| \leq \delta_{tol}$

Solution: Nothing will happen

1.5.2. Load Optimization

Formal description: $|t_i(j)-t_j(i)| \leq \gamma_{tol}$ and $|O_i - w_i| > \delta_{tol}$

Case (a): $w_i > O_i$ and $w_j < O_j$

Node i balances the workload by transferring some of its services to j, regardless of whether they are important or not since the trust of nodes is similar. Firstly, it determines $\Psi_{i,j}$ (see Formula 1.3 and 1.4) as a set of services that could be selected to balance the workload of nodes. Note that $C(I_s)$ represents the consumption function of a set of services I_s and is calculated by the sum of all its service consumptions.

$$\Psi_{i,j} = \{\, I_s | I_s \subseteq \left(S_i^{imp} \cup S_i^{unimp}\right) : maxC(I_s) \; and \qquad (1.3)$$

$$C\,(I_s) \leq \left(O_j - w_j\right) \; and \; 0 < C(I_s) \leq (w_i - O_i)\}$$

$$C(I_s) = \textstyle\sum_{s \in I_s} C_s \qquad (1.4)$$

If $\Psi_{i,j}$ is empty, then no optimization is done. Otherwise i transfers $\Psi_{i,j}$ to j.

Case (b): $w_i < O_i$ and $w_j > O_j$

Since services are assumed not to be stolen from other nodes, node i sends an alert message to j to piggy-back information necessary for self-optimization as described above. Then, case (1.5.2-a) will be triggered but on the side of j.

1.5.3. Trust Optimization

Formal description: $|t_i(j)-t_j(i)| > \gamma_{tol}$ and $|O_i - w_i| \leq \delta_{tol}$

Case (a): $t_j(i) > t_i(j)$

In this case i determines $\Psi_{i,j}$ (see Formula 1.5) as a set of unimportant services (i.e., with the maximum load consumption) that could be exchanged for important services of j so that the difference of their load consumption never exceeds C_{tol} to keep the loadbalancing property in both nodes satisfied.

$$\Psi_{i,j} = \{ I_s | I_s \subseteq S_i^{unimp}, \exists J_s \subseteq S_j^{imp} : maxC(I_s) \ and \quad (1.5)$$

$$|C(I_s) - C(J_s) \leq C_{tol} \ and \ (C(I_s) + w_i) \leq c_j^{max}\}$$

Then, after transferring $\Psi_{i,j}$, node i sends an alert optimization message to j (i.e., including all information which are necessary for the optimization) in order to trigger case (1.5.4-b) on side of j. Note that the execution of this step aims to balance again the workload between the nodes.

Case (b): $t_j(i) < t_i(j)$

In contrast to case (1.5.3-a), $\Psi_{i,j}$ is determined only from important services (see Formula 1.6), since j is more trustworthy than i. Then, i sends an alert optimization message to j in order to trigger case (1.5.4-a) on side of j.

$$\Psi_{i,j} = \{ I_s | I_s \subseteq S_i^{imp}, \exists J_s \subseteq S_j^{unimp} : maxC(I_s) \ and \quad (1.6)$$

$$|C(I_s) - C(J_s) \leq C_{tol} \ and \ (C(I_s) + w_i) \leq c_j^{max}\}$$

1.5.4. Trust and Load Optimization

Formal description: $|t_i(j)-t_j(i)| > \gamma_{tol}$ and $|O_i - w_i| > \delta_{tol}$

Case (a): $w_i > O_i$ and $w_j < O_j$ and $t_j(i) > t_i(j)$

Node i balances the workload only by transferring unimportant services to j (i.e., due to the fact that i is more trustworthy than j). It determines

$\Psi_{i,j}$ as a set of only unimportant services that could be selected to balance the workload of nodes (see Formula 1.7). Then, i transfers $\Psi_{i,j}$ to j.

$$\Psi_{i,j} = \{\ I_s | I_s \subseteq S_i^{unimp} : maxC(I_s) \qquad (1.7)$$

$$and\ C\ (I_s) \leq (O_j - w_j)\ and\ 0 < C(I_s) \leq (w_i - O_i)\}$$

Case (b): $w_i > O_i$ and $w_j < O_j$ and $t_j(i) < t_i(j)$

Since j is more trustworthy than i, $\Psi_{i,j}$ will be determined only from important services (see Formula 1.8). Then, just as the case of (1.5.4-a), if $\Psi_{i,j}$ is empty, no optimization is done. Otherwise i transfers $\Psi_{i,j}$ to j.

$$\Psi_{i,j} = \{\ I_s | I_s \subseteq S_i^{imp} : maxC(I_s) \qquad (1.8)$$

$$and\ C\ (I_s) \leq (O_j - w_j)\ and\ 0 < C(I_s) \leq (w_i - O_i)\}$$

In other cases:

Node i sends an alert message to j (i.e., including all information which are necessary for the optimization). Depending on the situation, case (1.5.4-a or 1.5.4-b) will then be triggered on the side of j.

1.6. Multiple Simultaneous Requests

In the evaluation, we have shown that the basic self-optimization algorithm presented in Section 1.5 led to good performance in terms of trust and workload, but we think that there is a room for improvement with the mechanism presented in this section. Therefore, we analyze now a network situation consisting of multiple simultaneous requests which are addressed to a single node to trigger the self-optimization process. Fig. 1.3 gives an overview of this situation. Let n_i denote the node that receives the requests and let be $L^i = \{l_1, l_2, ..., l_k\}$ the set of requesters considered by n_i. We first start with the description of the environment of n_i that has full information about its requesters. It can easily determine the set of potential service transfers Ψ_{n_i,l_j} for each requester $l_j \in L^i$, using the equations cited in Section 1.5, depending on the current situation of nodes. In the basic approach, as shown in Fig. 1.3, n_i optimizes itself with the requesters one after another in a random way without having preference for those that have many potential service transfers. By this means, the overall optimization in the system might take a long time before a large amount of services are transferred, particularity with a growing number of requesters. As a result, too much time can be spent in the whole system to get better optimized nodes. Our goal is to reduce this time by transferring the maximum amount of services as early as possible at runtime. Two approaches can be used to handle this problem.

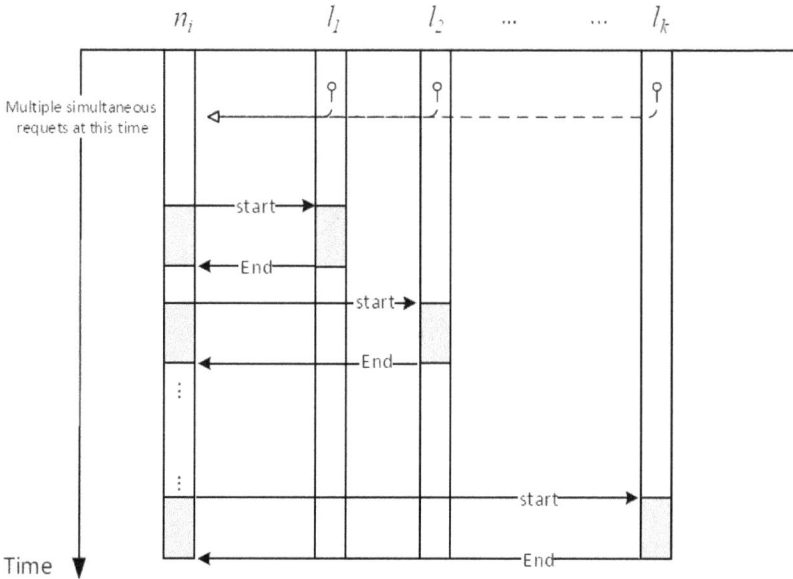

Fig. 1.3. Current execution of the basic algorithm.

1.6.1. Selective Request Handling

The first approach is called *selective request handling* because it always allows n_i to select the best requester to perform the optimization. We make use of two parameters in our approach, namely X and S_Ψ. The first parameter X is initialized as the set of all involved requesters — in our case always L^i — and S_Ψ is an empty list of fixed size $|L^i|$ used to store the potential number of service transfers. The basic idea behind the algorithm is: Whenever n_i receives multiples requests, it calculates the number of service transfers for every requester and applies an optimization with the requester whose services are most among the remaining requesters in X. If there is no requester with such a property, nothing will be done, as the nodes are already optimized. Otherwise, the found requester is removed and this process is repeated until all requesters are processed. In Algorithm 1, the above described algorithm is formalized as pseudo-code. This approach is very simple – and even in the worst case it is at least never worse than doing optimization with random selection – but the optimization output might be suboptimal regarding the overall self-optimization time due to its sequential processing. Therefore, we are interested in the second approach to provide a solution which supports parallelism through the optimization of requesters.

Algorithm 1. Node n_i:

1: $X \leftarrow L^i$;————→initialize X as the set of all involved requesters

2: S_Ψ = nil _____→ S_Ψ is initialized as empty list of fixed size $|L^i|$

3: **for** $x \in X$ **do**

4:　　　calculate $|\Psi_{n_i x}|$ and append it to S_Ψ

5: **end for**

6: **while** $X \neq \emptyset$ **do**

7:　　　select from S_Ψ the requester x with:

8:　　　$\{x | \exists x \in X : |\Psi_{n_i x}|$ is max and $|\Psi_{n_i x}| > 0\}$

9:　　　**if** no requester with such a property exists **then**

10:　　　___**exit**

11:　　　**else**

12:　　　___x perform an optimization with n_i

13:　　　___remove x from X

14:　　　**end if**

15: **end while**

1.6.2. Parallel Request Handling

While in the first approach we match n_i to a single requester to perform the optimization process, in this approach we consider a parallel optimization between requesters that work together to maximize the number of service transfers, as shown in Fig. 1.4. This has the benefit to further decrease the optimization time in the whole system. However, nodes in our system have different trust and workload values and some of them can transfer more services with one than others. Therefore, an important aspect for n_i is the formation of pairs between nodes — to apply the optimization algorithm in pairs and parallel — but in a way that the number of service transfers will be maximized in the system in order to deliver better results. Algorithm 2 shows the proposed mechanism formalized as pseudo-code.

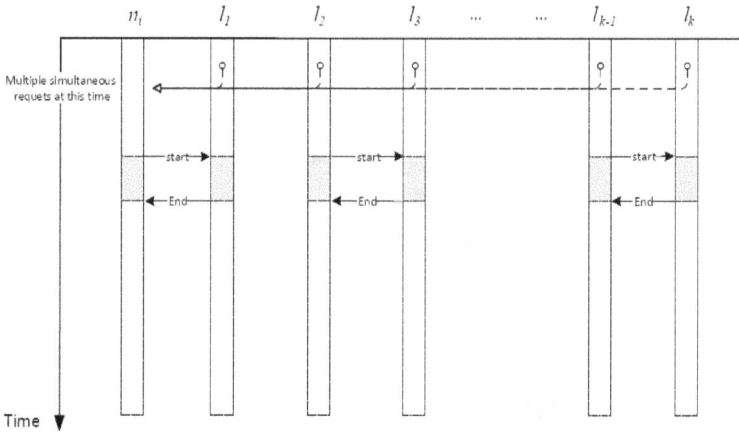

Fig. 1.4. Simplified representation of the parallel request handling.

At the beginning, we initialize two parameters X and T_Ψ. The first parameter $X = \{n_i\} \cup L^{\ i}$ represents the set of all nodes involved in the multiple requests, whereas the second parameter T_Ψ stands for an integer matrix of size $|X| \times |X|$, which we use to store the number of service transfers between nodes. Again, we say that x can optimize itself better with y than z, if and only if $|\Psi_{x,z}| \leq |\Psi_{x,y}|$ with $y \neq z$. Then, the algorithm is split into two phases, the first of which is similar to the selective request handling, but we now allow to calculate the number of service transfers between any two nodes in X. Intuitively, reflexive suitability values such as $\Psi_{x,x}$ are not computable in this phase, simply because it is not allowed that a node is optimizing itself. Afterwards, the algorithm enters in its second phase exploring pairs having at least a service transfer of one and maximizing at the same time the number of service transfers. If there is no pair with such a property, the algorithm terminates. Otherwise, the found pair becomes engaged to perform the optimization process. Then, the pair is finally removed from the set of X. The while loop continues until there are no more pairs to perform the optimization process. To demonstrate the proposed algorithm an example is discussed.

Algorithm 2 Node n_i:

1: $X \leftarrow \{n_i\} \cup L^{\,i}$ → initialize X as the set of all involved nodes

2: $T_\Psi \leftarrow$

	n_i	l_1	...	l_k
n_i	0	0	...	0
l_1		0	...	0
...			0	0
l_k				0

→ is an empty lookup table of size $|X| \times |X|$

Phase 1
3: **for** $x \in X$ **do**
4:     **for** $y \in X \setminus \{x\}$ **do**
5:         calculate $|\Psi x,y|$ and append it to T_Ψ
6:     **end for**
7: **end for**

Phase 2
8: **while** two nodes remain in X **do**
9:     select from T_Ψ the pair (x,y) with:
10:    $\{(x,y)|\exists x,y \in X : |\Psi_{x,y}| \text{ is max and } |\Psi x,y| > 0\}$
11:     **if** no pair with such a property exists **then**
12:         **exit**
13:     **else**
14:         x and y become engaged to perform the optimization
15:         remove x and y from X
16:     **end if**
17: **end while**

Example: In this example, an instance of parallel request handling involving five requesters is considered, with $X = \{n_i, l_1, l_2, l_3, l_4, l_5\}$. We assume that the set of service transfers between nodes has already been processed by n_i, leading to the relation graph illustrated in Fig. 1.5.

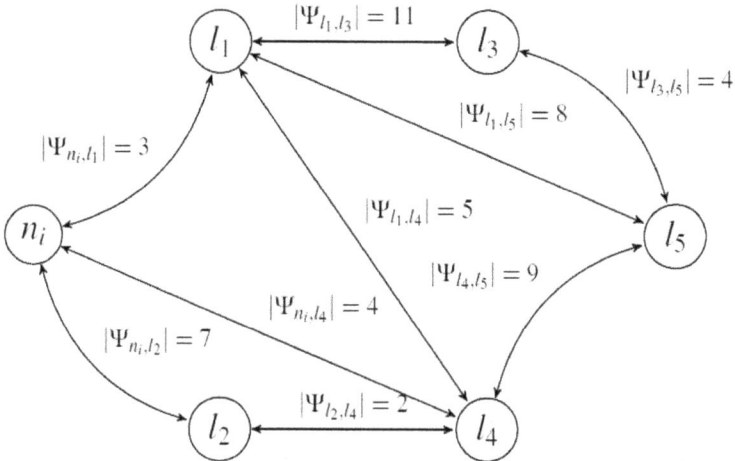

$$|\Psi_{l_1,l_3}| = 11$$
$$|\Psi_{l_3,l_5}| = 4$$
$$|\Psi_{n_i,l_1}| = 3$$
$$|\Psi_{l_1,l_5}| = 8$$
$$|\Psi_{l_1,l_4}| = 5$$
$$|\Psi_{l_4,l_5}| = 9$$
$$|\Psi_{n_i,l_4}| = 4$$
$$|\Psi_{n_i,l_2}| = 7$$
$$|\Psi_{l_2,l_4}| = 2$$

Fig. 1.5. Relation graph of potential service transfers.

Based on this information, the algorithm starts its first phase by calculating T_Ψ. So phase one ends with the table of matrix presented in Fig. 1.6.

$$T_\Psi =$$

	n_i	l_1	l_2	l_3	l_4	l_5
n_i		3	7	0	4	0
l_1			0	11	5	8
l_2				0	2	0
l_3					0	4
l_4						9
l_5						

Fig. 1.6. A simplified representation of T_Ψ after the execution of phase one.

In the second phase, we need to define for each node its best partner that contributes to maximize the service transfers in the whole system. In the iteration $loop_1$ the pair (l_1,l_3) is identified first. This is because (l_1,l_3) returns the maximum number of service transfers in T_Ψ. Eliminating them gives $X = \{n_i, l_2, l_4, l_5\}$. Next, pair (l_4, l_5) is identified in *loop$_2$* and its elimination yields $X = \{n_i, l_2\}$. Finally, the pair (n_i, l_2) is identified and its elimination gives $X = \{0/\}$. Hence, the algorithm finishes with the following optimization pairs $\{(l_1, l_3), (l_4, l_5), (n_i, l_2)\}$.

1.7. Evaluation

In this section an evaluation for the introduced self-optimization approach is provided. For the purpose of evaluating and testing, an evaluator based on the TEM middleware [11] has been implemented which is able to simulate the self-optimization algorithm. The evaluation network consists of 100 nodes, where all nodes are able to communicate with each other using message passing. Experiments with more nodes were tested and yielded similar results, but with 100 nodes more observable effects were seen. Each node has a limited resource capacity (memory) and is judged by an individual trust value without any central knowledge. Furthermore, four type of nodes are defined with different trust and resource values (see Table 1.1).

Table 1.1. Mixture of heterogeneous nodes.

Node Type	Memory (MB)	Trust	Amount (%)
Type 1	[500 - 1000]	[0.7 - 0.9]	10
Type 2	[500 - 1500]	[0.3 - 0.6]	50
Type 3	[2000 - 4000]	[0.4 - 0.8]	30
Type 4	[4000 - 8000]	[0.4 - 0.9]	10

Then, a mixture of heterogeneous services with different resource consumptions are randomly generated for nodes. The sum of all node's service consumptions does not exceed a node's capacity (i.e., as defined in Formula 1.1). If, for example, a trustworthy node is already full, then the same procedure is repeated for an untrustworthy node and so on until the average load of the system reaches 50 % ($\overline{workload} = 50\,\%$). This means that some nodes may have many services and others none to unbalance the workload between nodes. Important services are created only for untrustworthy nodes and unimportant services for trustworthy nodes. Without the self-optimization techniques the workload of nodes are still unbalanced. Moreover, important services running on untrustworthy nodes are prone to fail. With the use of direct trust and reputation, the trust of a node can be measured and taken into consideration for the transfer of services. Two rating functions are used to evaluate the fitness of a service distribution regarding trust and workload. The first rating function for workload $F_{workload}$ aims to calculate the average deviation of all nodes from the desired workload $\overline{workload}$ (in our case, 50 %). This is expressed by the Formula 1.9, where N is the set of all nodes and $|N|$ the cardinality of N. The main idea of the second rating function F_{trust} is to reward important services running on trustworthy nodes. This is expressed by the Formula 1.11, where N is the set of all nodes, S_n is the set of services on a node n, $t(n)$ its trust value and p(s) the priority of a service s (i.e., if s is important, $P(s)$ has the value of 1, otherwise 0).

$$F_{workload} = \frac{\sum_{n \in N} |workload(n) - \overline{workload}|}{|N|} \qquad (1.9)$$

$$\overline{workload} = \frac{\sum_{n \in N} workload(n)}{|N|} \qquad (1.10)$$

At the beginning of the simulation, the network is rated by using both F_{trust} and $F_{workload}$. Then, the simulation is started and after each optimization step the network is rated again. Within one optimization step, 50 pairs of nodes (sender/receiver) are randomly chosen to perform the self-optimization process, i.e., $\rho = 50$ %. Senders send an application message to receivers to piggyback necessary information for the self-optimization, as described in Section 1.3. Based on the extracted information the receiver determines whether it transfers its services or not. The goal is to maximize the availability of important services, which means that F_{trust} should be maximized (i.e., to an optimal theoretical point that we explain later in 1.7.2). Therefore, it is necessary to transfer the more important services to more trustworthy nodes. Furthermore, the overall utilization of resources in the network should be well-balanced, i.e., $F_{workload}$ should be minimized near to zero.

$$F_{trust} = \sum_{n \in N} \sum_{s \in S_n} p(s)t(n) \qquad (1.11)$$

1.7.1. Results Regarding the Rating Function $F_{workload}$

As mentioned above, the first rating function $F_{workload}$ indicates the average workload deviation of all nodes from the desired workload

workload (in our case, 50 %). The lower the value of $F_{workload}$, the better the performance of workload balancing.

Fig. 1.7 shows the result of this experiment, whereas the values on the x-axis stand for optimization steps and the average workload deviation of nodes is depicted on the y-axis. It can be observed that the proposed algorithm improves the workload balancing by about 93 %. However, it does not reach the theoretical maximum rate of 100 % due to the trade-off between trust and workload.

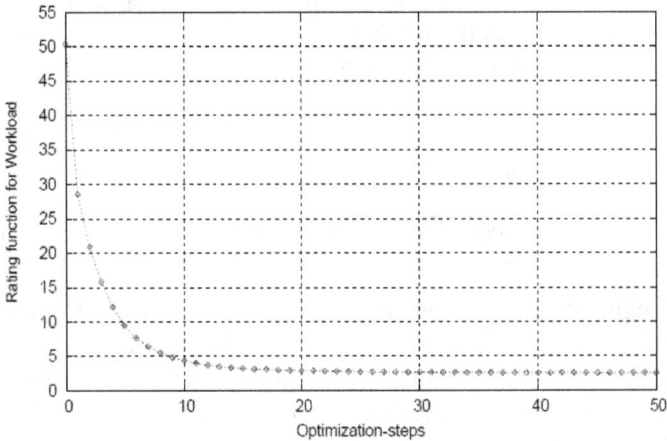

Fig. 1.7. Rating function for the workload deviation ($F_{workload}$).

1.7.2. Results Regarding the Rating Function F_{trust}

In the following, the service distribution for the proposed self-optimization algorithm is evaluated regarding F_{trust}.

Fig. 1.8 shows the result of this experiment. The square line represents the result of F_{trust} using the proposed self-optimization algorithm. It can be observed that the algorithm improves during runtime the availability of important services. This means that the consideration of workload does not prevent the algorithm to relocate important services to trustworthy nodes. However, it remains to investigate the quality of the obtained result compared to an optimal theoretical result, when all important services are hosted only on trustworthy nodes (pure trust distribution, i.e., regardless of whether nodes are balanced or not). For this purpose we use an approximation algorithm that sorts in decreasing order the trust values of nodes and relocates all important services only to most trustworthy nodes until their capacity is full. The triangular marked line in the figure illustrates the result of the approximation algorithm. As a conclusion to all simulations we have done so far (about 1000 runs were evaluated) we can state that the proposed algorithm greatly improves the trust distribution of services. More precisely, it achieves 85 % of the theoretical maximum result. However, it stays by 15 % behind the theoretical maximum result due to the trade-off between trust and workload.

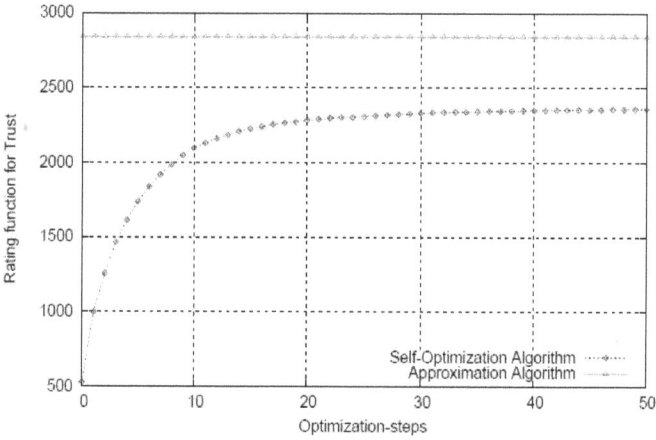

Fig. 1.8. Rating function for Trust (F_{trust}).

1.7.3. Basic Algorithm vs. Extensions

In this section, the gain of applying the proposed extensions with respect to Section 1.6 is investigated. We use the similar parameter settings of the initial evaluation, but we now allow for a certain percentage of randomly chosen nodes to receive multiple optimization requests simultaneously. This has the benefit to put the evaluation more in a context of real life. In this part of work, the following three algorithms are compared regarding their ability to perform the optimization in the system.

- **Basic algorithm (ALG.1):** The basic optimization algorithm as in the previous experiments.

- **Basic algorithm + Selective Request Handling (ALG.2):** A variation of the basic optimization algorithm using the extension of the selective request handling (see Section 1.6.1).

- **Basic algorithm + Parallel Request Handling (ALG.3):** A variation of the basic optimization algorithm using the extension of the parallel request handling (see Section 1.6.2)

The three algorithms differ in the way they handle multiple requests, either sequential or parallel. Figs. 1.9 and 1.10 present their comparison results with respect to the rating functions F_{trust} and $F_{workload}$.

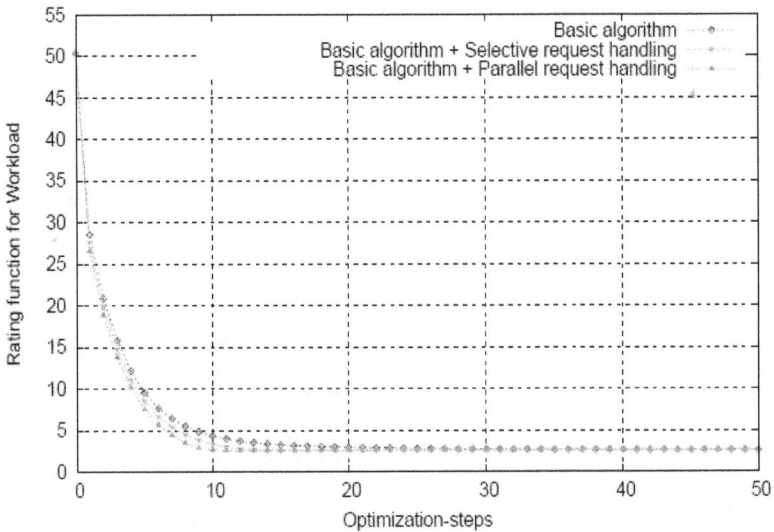

Fig. 1.9. Comparison results according to the rating function $F_{workload}$.

It is easy to see that both investigated variations of ALG.2 and ALG.3 indeed provide an even better optimization time than the basic algorithm ALG.1, especially the variation of ALG.3, currently shows the best time performance to achieve the optimization process. This is due to its ability to support parallelism through the optimization of requesters such that everyone optimizes itself with the node with the highest gain of service transfers. This results - in the whole system - to a reduce of the processing time into the overall optimization.

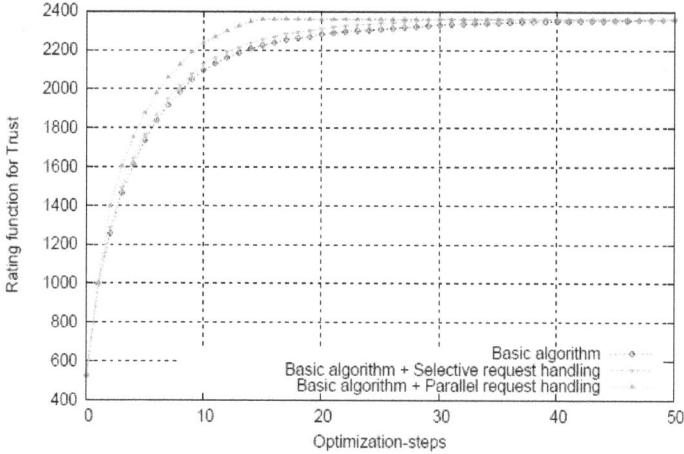

Fig. 1.10. Comparison results according to the rating function F_{trust}.

1.7.4. Different Network Settings

In the following, additional experiments are conducted to further investigate the behavior of the introduced self-optimization algorithm with different network settings. We performed a binary classification of nodes with a ratio of 50/50, and for each classification type, we generated a different amount of memory resources and trust values, as shown in Table 1.2. Generally, the more trustworthy the nodes are, the higher is the amount of their memory resources. We argue that this is a useful and realistic network parametrization since it enables to model the behaviour of servers and workstations which are expected to be trustworthy in real-world situations through the use of Type 1 as well the behavior of mobile devices (i.e., expected in real-world to be less trustworthy than servers and workstations) through the use of Type 2. The average workload is set to 45 %. The experiments differ in the adjustment of $|N|$ and ρ. Recall, $|N|$ states for the size of the network and ρ represents the percentage amount of involved nodes within one optimization step to perform the optimization process. In the following the results of conducted experiments are presented. To ensure representative values, any experiment is repeated 300 times and the results are averaged.

The first three experiments examine the behaviour of the self-optimization algorithm with a fixed $|N|$ = 100 but different percentage of ρ.

Table 1.2. A binary classification of heterogeneous nodes.

zNode Type	Memory (MB)	Trust	Amount (%)
Type 1	[8000 - 16000]	[0.6 - 0.99]	50
Type 2	[1000 - 8000]	[0.1 - 0.60]	50

• Experiment 1.1: $|N| = 100$, $\rho = 30\%$ (see Figs. 1.11 and 1.12)

• Experiment 1.2: $|N| = 100$, $\rho = 50\%$ (see Figs. 1.11 and 1.12)

• Experiment 1.3: $|N| = 100$, $\rho = 70\%$ (see Figs. 1.11 and 1.12)

Experiments 2.1-2.3 consider a fixed network size of $|N| = 200$ and different percentage of ρ.

• Experiment 2.1: $|N| = 200$, $\rho = 30\%$ (see Figs. 1.13 and 1.14)

• Experiment 2.2: $|N| = 200$, $\rho = 50\%$ (see Figs. 1.13 and 1.14)

• Experiment 2.3: $|N| = 200$, $\rho = 70\%$ (see Figs. 1.13 and 1.14)

The following three experiments are similar to the first ones but the network size is set to $|N| = 400$.

• Experiment 3.1: $|N| = 400$, $\rho = 30\%$ (see Figs. 1.15 and 1.16)

• Experiment 3.2: $|N| = 400$, $\rho = 50\%$ (see Figs. 1.15 and 1.16)

• Experiment 3.3: $|N| = 400$, $\rho = 70\%$ (see Figs. 1.15 and 1.16)

The last three experiments examine the behaviour of the introduced algorithm with $|N| = 800$ and different ρ.

• Experiment 4.1: $|N| = 800$, $\rho = 30\%$ (see Figs. 1.17 and 1.18)

• Experiment 4.2: $|N| = 800$, $\rho = 50\%$ (see Figs. 1.17 and 1.18)

• Experiment 4.3: $|N| = 800$, $\rho = 70\%$ (see Figs. 1.17 and 1.18)

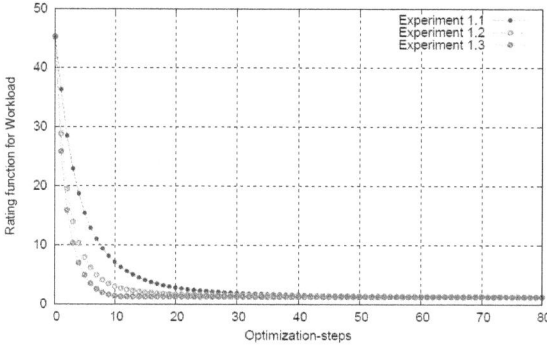

Fig. 1.11. Result of experiments 1.1 - 1-3 according to the rating function $F_{workload}$.

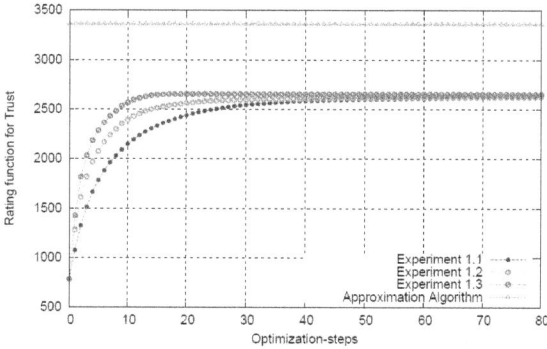

Fig. 1.12. Result of experiments 1.1 - 1-3 according to the rating function F_{trust}.

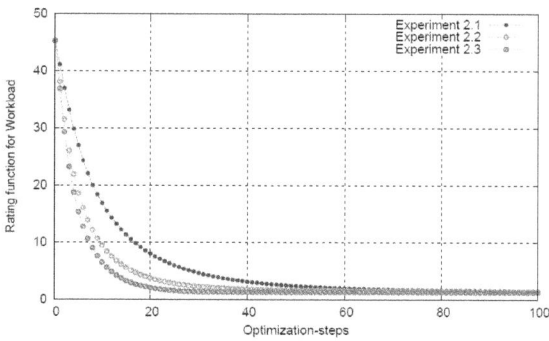

Fig. 1.13. Result of experiments 2.1 - 2-3 according to the rating function $F_{workload}$.

41

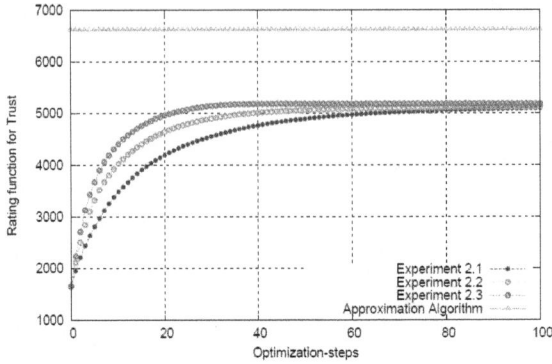

Fig. 1.14. Result of experiments 2.1 - 2-3 according to the rating function F_{trust}.

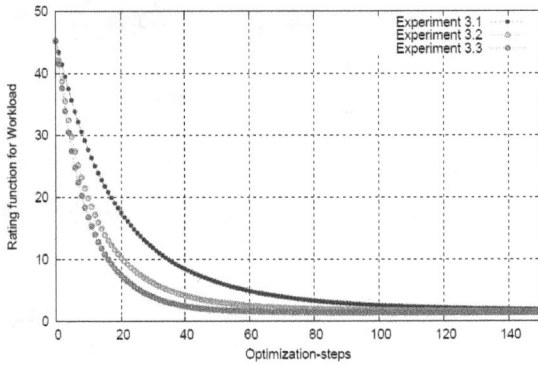

Fig. 1.15. Result of experiments 3.1 - 3-3 according to the rating function $F_{workload}$.

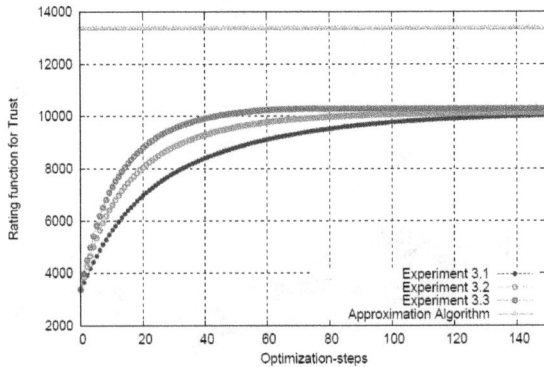

Fig. 1.16. Result of experiments 3.1 - 3-3 according to the rating function F_{trust}.

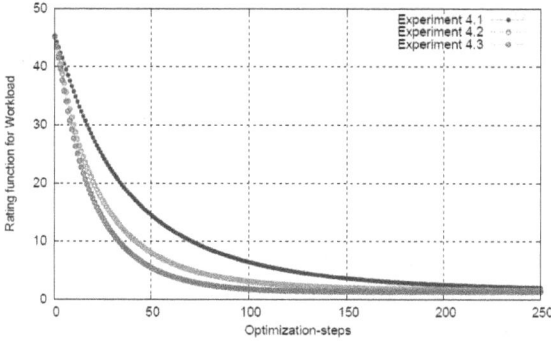

Fig. 1.17. Result of experiments 4.1 - 4-3 according to the rating function $F_{workload}$.

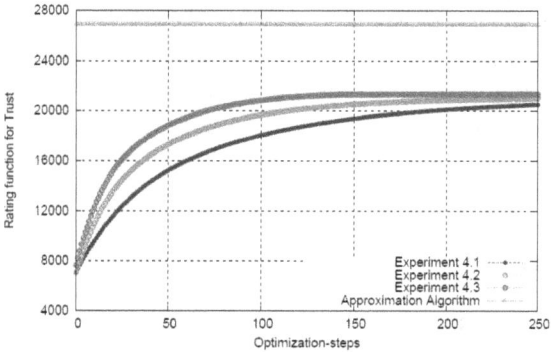

Fig. 1.18. Result of experiments 4.1 - 4-3 according to the rating function F_{trust}

Conclusion Deduced From Conducting Experiments. The experiment results, with the focus on workload, are depicted in Figs. 1.11, 1.13, 1.15, and 1.17. These figures show the optimization steps on the horizontal axis and the workload deviation of nodes on the vertical axis. Values near to the bottom left corner represent small deviation of workloads with few number of optimization steps. The results attest the introduced self-optimization algorithm a continuous reduction of the workload deviations in all kind of settings. Beside the workload balancing, the introduced algorithm provides also a good ability to improve its speedup over the parametrization of ρ, making it suitable to be applied in overfilled situations with too many number of messages. Figs. 1.12, 1.14, 1.16, and 1.18 show similar results to the workload experiments, but with the focus on trust. The optimization steps are depicted on the horizontal axis and the fitness function for trust on the vertical axis. Optimal theoretical values considering pure trust distributions are marked with red triangular lines for each experiment. Similarly to the last results, we can state that the algorithm developed in this work is able to always improve the availability of important services at runtime and that the parametrization of ρ plays here also an important role to increase the speedup of the trust optimization in the whole system.

1.8. Conclusions and Future Work

In this chapter, a novel self-optimization algorithm for open distributed self-* systems has been proposed. The algorithm does not only consider pure load-balancing but also takes into account trust to improve the assignment of important services to trustworthy nodes at runtime. More precisely, the algorithm makes use of different optimization strategies — - as cited in the corresponding part of Section 1.5 — to determine whether a service should be transferred to another node or not. Section 1.7 presents the results of the performance measurements that are conducted to evaluate the algorithm. The results indicate that for our model trust concepts improve significantly the availability of important services while causing a small deterioration (i.e., by about 7 %) regarding load balancing. Therefore, we classify our algorithm as a kind of best-effort approach that provides good but not necessarily optimal solutions to this trade-off problem. Then, a set of variations of the basic algorithm are introduced in Section 1.6 to improve its performance in case of multiple requests. The difference between the variations arises in the way to handle requests, either sequential or parallel. In Section 1.7.3, a comparative evaluation is conducted to analyze the performance results of the variations compared to the basic approach. The results attest a good performance for the extended optimization algorithm with parallel request handling. In Section 1.7.4, an additional evaluation is provided to further investigate the behavior of our approach for different network settings. The results indicate here as well a good performance for our algorithm. It clearly attains its goals of both trust and load optimizations in all kind of parametrizations and network sizes. Apart from this, the algorithm provides also a good possibility to increase its speedup over the parametrization of ρ making it suitable to be applied in overfilled situations with too many number of messages. In future work, extensions are planned to deal with the Cold-Start-Problem, i.e., the need to integrate new nodes with unknown trust values with other nodes in the network. This is very important to improve the robustness of the proposed self-optimization algorithm. One possible solution to address this issue could be to make runtime prediction or online training for the new participating nodes, but as it goes beyond the scope of this work it is not further discussed here.

References

[1]. Msadek, N., Kiefhaber, R., Fechner, B., Ungerer, T., Trust-enhanced self-configuration for organic computing systems, in *Proceedings of the 27th International Conference on Architecture of Computing Systems (ARCS2014)*, 2014.

[2]. Msadek, N., Kiefhaber, R., Ungerer, T., Simultaneous self-configuration with multiple managers for organic computing systems, in *Proceedings of the 2nd International Workshop on Self-optimisation in Organic and Autonomic Computing Systems (SAOS'14) in conjunction with ARCS' 14*, 2014.

[3]. Msadek, N., Kiefhaber, R., Ungerer, T., A trustworthy, fault-tolerant and scalable self-configuration algorithm for organic computing systems. *Journal of Systems Architecture (JSA)*, Vol. 61, 2015, pp. 511 – 519.

[4]. Msadek, N., Kiefhaber, R., Ungerer, T., Trustworthy self-optimization in organic computing environments, in *Proceedings of the 28th International Conference on Architecture of Computing Systems Series*, Vol. 9017, 2015, pp. 123–134.

[5]. Msadek, N., Kiefhaber, R., Ungerer, T., A trust- and load-based self-optimization algorithm for organic computing systems, in *Proceedings of the International Conference on Self-Adaptive and Self-Organizing Systems (SASO)*, 2014.

[6]. Msadek, N., Ungerer, T., Trustworthy self-optimization for organic computing environments using multiple simultaneous requests, *Journal of Systems Architecture (JSA)*, Vol. 75, 2017, pp. 26 –34.

[7]. Msadek, N., Ungerer, T., Trust-based monitoring for self-healing of distributed real-time systems, in *Proceedings of the 7th IEEE Workshop on Self-Organizing Real-Time Systems (SORT'16) in conjunction with ISORC'16*, 2016.

[8]. Msadek, N., Ungerer, T., Trust as important factor for building robust self-x systems, in *Proceedings of the Trustworthy Open Self-Organising Systems*, 2016.

[9]. Msadek, N., Trust as a principal ingredient to improve the robustness of self-organizing systems, in *Proceedings of the Organic Computing: Doctoral Dissertation Colloquium*, 2015.

[10]. Kiefhaber, R., Jahr, R., Msadek, N., Ungerer, T., Ranking of direct trust, confidence, and reputation in an abstract system with unreliable components, in *Proceedings of the 10th IEEE International Conference on Autonomic and Trusted Computing (ATC-13)*, 2013.

[11]. Anders, G., Siefert, F., Msadek, N., Kiefhaber, R., Kosak, O., Reif, W., Ungerer, T., Temas a trust-enabling multi-agent system for open environments. Technical report, *Universitat Augsburg*, 2013.

[12]. Khayyat, Z., Awara, K., Alonazi, A., Jamjoom, H., Williams, D., Kalnis, P., Mizan: a system for dynamic load balancing in large-scale graph processing, in *Proceedings of the 8th ACM European Conference on Computer Systems,* 2013, pp. 169–182.

[13]. Babak, H., Kit, L. Y., Lilja, D. J., Dynamic task scheduling using online optimization, in *Proceedings of the Journal IEEE Transactions on Parallel and Distributed Systems.,* Vol. 11, Issue 11, 2000.

[14]. Panwar, R., Mallick, B., Load balancing in cloud computing using dynamic load management algorithm, in *Proceedings of the International Conference on Green Computing and Internet of Things (ICGCIoT),* 2015, pp. 773–778.

[15]. Siar, H., Kiani, K., Chronopoulos, A. T., An effective game theoretic static load balancing applied to distributed computing, *Journal on Cluster Computing,* Vol. 18, Issue 4, 2015, pp. 1609–1623.

[16]. Anis Uddin Nasir, M., De Francisci Morales, G., Garcia-Soriano, D., Kourtellis, N., Serafini, M., The power of both choices: Practical load balancing for distributed stream processing engines, in *Proceedings of the IEEE 31st International Conference on Data Engineering (ICDE' 15),* 2015, pp. 137–148.

[17]. Akbar, A., Basha, S. M., Abdul Sattar, S., A comparative study on load balancing algorithms for sip servers, *Information Systems Design and Intelligent Applications Series,* Vol. 435, 2016, pp. 79–88.

[18]. Rao, A., Lakshminarayanan, K., Surana, S., Karp, R., Stoica, I., Load balancing in structured p2p systems, in *Proceedings of the 2nd International Workshop (IPTPS),* 2012.

[19]. Bittencourt, L., Madeira, E. R. M., Cicerre, F. R. L., Buzato, L. E., A path clustering heuristic for scheduling task graphs onto a grid, in *Proceedings of the 3rd International Workshop on Middleware for Grid Computing (MGC05),* 2005.

[20]. Lobinger, A., Stefanski, S., Jansen, T., Balan, I., Coordinating handover parameter optimization and load balancing in the self-optimizing networks, in *Proceedings of the IEEE 73rd Vehicular Technology Conference (VTC Spring),* 2011.

[21]. Wang, K., Zhou, X., Li, T., Zhao, D., Lang, M., Raicu, I., Optimizing load balancing and data-locality with data-aware scheduling, in *Proceedings of the IEEE International Conference on Big Data,* 2014, pp. 119–128.

[22]. Satzger, B., Mutschelknaus, F., Bagci, F., Kluge, F., Ungerer, T., Towards trustworthy self-optimization for distributed systems, in *Proceedings of the Software Technologies for Embedded and Ubiquitous Systems Lecture Notes in Computer Science,* Vol. 5860, 2009, pp. 58-68.

[23]. Derek L, E., Edward, D. L., John, Z., A comparison of receiver-initiated and sender-initiated adaptive load sharing, in *Proceedings of the Conference on Measurement and Modeling of Computer Systems,* 1986.

2.

Task Mapping in Heterogeneous NoC by Means of Population-Based Incremental Learning

Freddy Bolanos, Fredy Rivera, Jose Aedo, Nader Bagherzadeh

2.1. Introduction

Speed and complexity of high-performance embedded systems has experienced an exponential increase, as a consequence of the improvements performed to the related manufacturing processes. For several years, Moore's law had predicted the speedup of computing systems based on one single core with some accuracy, by setting that such systems performance would double about every eighteen months. Since manufacturing processes for single core systems have experienced a saturation problem, it is no longer possible to improve the capacities of such systems by simply taking advantage of lower scales of integration. Instead of that, *multicore* and *manycore* systems seems to provide a suitable tradeoff between performance and system complexity, at expenses of adding more and more Processing Elements (PE) to the system architecture. Given these conditions, Moore's law has been revisited, and now it is predicted an exponential growing in the amount of PEs present in the embedded system, in order to offering higher performances and coping the market requirements [1].

By leaving behind the single-core paradigm, some new issues appear, related mainly with the communication among the several PEs. This feature becomes a bottleneck in embedded systems design, since it increasingly affects both performance and power consumption of multicore and manycore systems. Such issues have increased in complexity as the amount of PEs or system cores rises, and have

Freddy Bolanos
Electrical Engineering and Automatics Department Universidad Nacional de Colombia, Colombia

triggered the throughput gap for the design process. Indeed, nowadays a designer has the availability of an increased amount of processing resources, which are often heterogeneous in nature (i.e. there are several types of PE, with diverse features), and must able to perform communication tasks among each others. As a consequence, the evolving rates for resources availability and design throughput are quite different, as depicted in Fig. 2.1. This situation is referred to as the Design Productivity Gap, which is nothing but the inability of current design teams for facing complex projects, without an overshoot in time or resources [2].

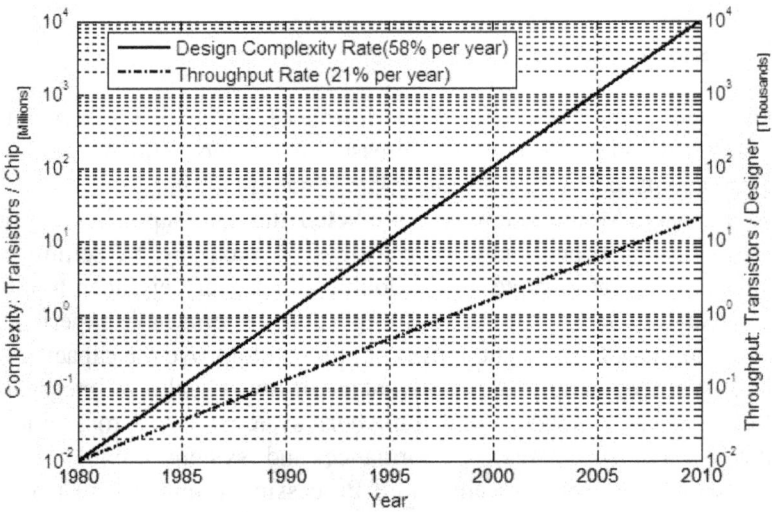

Fig. 2.1. The Design Productivity Gap [2].

Network on Chip (NoC) architectures have been proposed for the sake of providing scalability to multicore and manycore systems, while meeting performance, energy and several other constraints. A NoC architecture is composed by a set of cores or PEs, as well as a set of communication resources, often organized as a regular network topology, as depicted in Fig. 2.2.

The NoC depicted in Fig. 2.2 has a set of nine cores, organized in a two-dimensional mesh. Each core or PE is connected to the communication network through a Network Interface (NI in the figure), which implements the whole features needed to send and receive data to and from the network. Each Network Interface is in turn connected to a node router, which implements algorithms for guaranteeing that information may travel dependably from source to destination nodes. Finally, the network is completed by means of high speed links between pairs of routers, which transport the data among PEs. Several efforts have been reported for improving the capabilities of NoC Architectures. Some of them are related to the communication hardware [3-5], and some others are devoted to improving the architecture configuration, the architecture topology, and the architecture parameters [6-9].

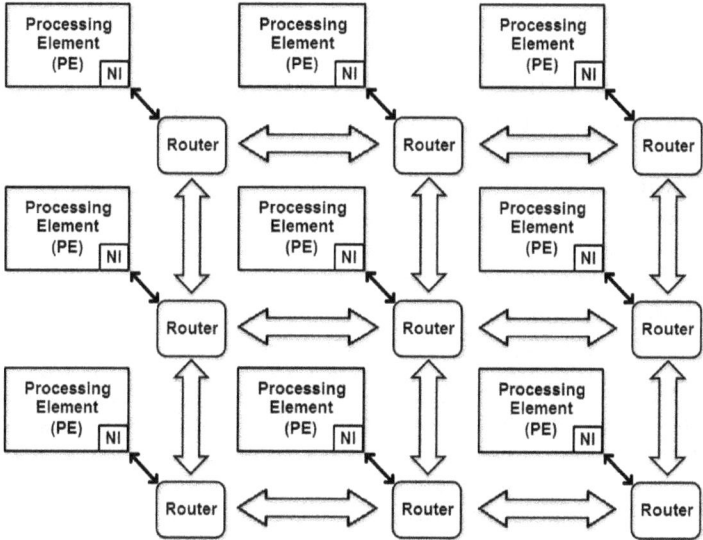

Fig. 2.2. A typical NoC, with 2D mesh topology.

Wireless RF links has been incorporated to the NoC systems, in order to improve the latency of communication links and the size of the network, without increasing prohibitively the power consumption [10-12]. These systems are referred to as Wireless NoC, WiNoC, or WNoC systems. The best tradeoff in these systems seems to be the use traditional wired links for short-range communications, and wireless channels for large-range links. That is why the systems which embodies both wire and wireless links on a single chip, are also referred as Hierarchical NoC systems.

The task mapping problem in NoC environments

As mentioned earlier, task mapping has become a very complex process in current embedded systems design flow. The NoC approach has allowed some degree of scalability, and to divide the software applications for exploding task parallelism and concurrency. However, a big issue appears in such conditions: Which is the best combination of PEs that must execute a given set of task from an application, optimizing in turn several figures of merit?

Task mapping in NoC environments depends both on the input applications as on the target architecture, and it is influenced by factors as diverse as platform and software constraints, optimization criteria, the design-space search engine and its limitations, and the available mapping information at design time [13]. Consequently, the task mapping problem in NoC systems is considered as a NP-complete problem.

The starting point of a task mapping specification, is a task graph AG, which may be unequivocally defined as:

$$AG = (T,D), \qquad (2.1)$$

where $T = \{t_1, t_2, ..., t_N\}$ represents the set of executable tasks to be mapped to the NoC Architecture, and $D = \{d_1, d_2, ..., d_j\}$ represents the set of data dependences among the tasks. As can be seen, the sets T and L on Equation (2.1) are both finites, of sizes N and j respectively, and both may contain information regarding constraints such as real time deadlines for a task, or maximum allowed latency for a data transference between two given tasks.

In the same way, the target architecture for a task mapping problem may be represented by means of a graph, called ArchG, as depicted following:

$$ArchG = (H,L), \qquad (2.2)$$

where $H = \{h_1, h_2, h_3, ..., h_M\}$ represents the set of PEs or nodes available for tasks execution, and $L = \{l_1, l_2, ..., l_k\}$ represents the set of links available on the NoC platform. Again, both M and k are finite, since the available hardware for implementing the NoC is constrained. In some cases, and depending on the nature of the communication links (i.e. wired, high speed, wireless, optical), the set L may be divided in subsets or hierarchies. Both sets (H and L) must contain profiling information that allows the estimation of figures of merit (power, bandwidth, execution time, and so on) for the sake of guiding the design space searching process.

A task mapping specification in NoC systems, must combine the information from the high level (i.e. the application graph AG in Equation (1)) with platform information from the hardware level (i.e. the architecture graph $ArchG$ in Equation (2)). Such a combination is often referred to as Common Domain Semantic [14], and it is frequently that takes the form of an Acyclic Directed Annotated Graph (ADAG). Fig. 2.3 depicts an instance of an ADAG with $N = 5$ and $j = 5$, according with Equation (1). In such figure, tasks are represented by vertices in the graph, while dependencies or links among tasks are represented by the edges. Annotations provide information about the potential implementation of each given task or dependence in the available resources, and serves for guiding the optimization process, supplying a way for comparing several mapping solutions. Such information may include, but is not restricted to, power consumption, execution time, bandwidth and any other result of implementing tasks or links on available resources for the target NoC architecture.

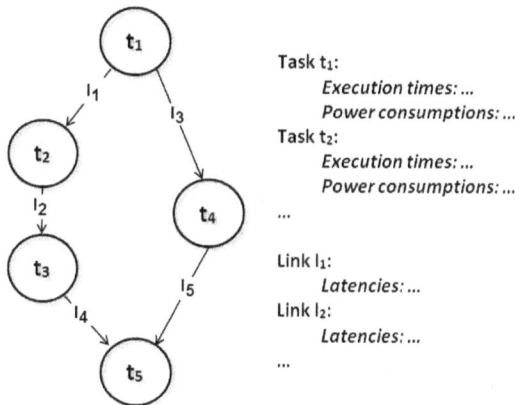

Fig. 2.3. An instance of an Acyclic Directed Annotated Graph.

Several task mapping automation proposals may be found in literature for NoC architectures. Depending on the figures of merit that are meant to optimize, the optimization engine may be classified as single-objective or multiobjective. Among the single-objective proposals, it can be found those devoted to optimize power consumption [15-17], and those which are meant to minimize the communication overheads [18-20]. Multiobjective approaches often combine figures of merit such as power consumption and communication costs [21, 22], some others are devoted to minimizing execution time and network bandwidth simultaneously [14], or minimizing various communication metrics at once [23, 24].

Another classification criterion for task mapping approaches reported in literature, is related to the target architecture. Some reported solutions are focused in homogeneous NoC target architectures [23-25, 17], which means that all nodes in set $H = \{h_1, h_2, ..., h_M\}$ of Equation (2) are identical. On the other hand, if nodes in set H may be different from each other, the target NoC architecture is said to be heterogeneous in nature. The reported solutions on this subject [26-28, 20] must deal with a more complex problem, since they must take into account the variability of the PEs features in the task mapping optimization. Finally, regarding the network topology, all the solutions listed so far are aimed at 2 dimensions meshes, or customized topologies.

A third criterion for classifying mapping solutions relies on the optimization engine, used to perform the design space exploration. As mentioned before, task mapping on NoC architectures is a very complex problem, so deterministic or exact optimization approaches are prone to worsen convergence time as the problem size rises. Similarly, given the complexity of the design space, those heuristic approaches which do not perform the optimization search in parallel (i.e. by exploring several potential solutions at once) are more prone to exhibit the local minima problem. As a consequence, some of the reported task mapping solutions correspond to heuristic algorithms, based on population [20, 29, 30].

2.2. The Population-Based Incremental Learning (PBIL) Algorithm

The PBIL approach is based on Genetic Algorithms (GA), and Competitive Learning (CL) Neural Networks. As derived from its name, the PBIL approach is based on a population of solutions, which means that it can perform a parallel searching in the solutions space. Such a parallel feature allows the algorithm avoiding the local optimum problem, and has the potential of speeding up the convergence time of the searching process [31]. As it does a GA, the PBIL algorithm uses a set of potential solutions (i.e. a population) to explore in several points of the searching space at once. The searching history (i.e. the learning information obtained from the exploration process) is stored in the form of a probabilities array. Such array may take several forms, depending on the optimization problem to be solved, ranging from probabilities vectors (one dimension) to two or three dimension matrices [32]. Fig. 2.4 depicts a typical PBIL probability matrix of two dimensions.

In Fig. 2.4, let us suppose a combinatorial optimization problem formed by N attributes (labeled as *Attribute* 1, *Attribute* 2, ..., *Attribute N*, in Fig. 2.4). These attributes are related to the optimization questions to be defined in order to obtain a suitable solution. For each of such attributes there may up to M potential choices, in such a way that a problem's solution is completely defined by $s_x = \{x_1, x_2, ... x_N\}$, where $1 \le x_j \le M$, $\forall$ $1 \le j \le N$. Each probability in the array depicted in Fig. 2.4, represents the potential for a single attribute to be solved using a given available choice. That is why the P matrix on figure has dimensions $M \times N$, and implies that a given probability $P(i,j)$ represents the probability of Attribute j, to be solved by using the respective choice i.

	Attribute 1	Attribute 2	...	Attribute j	...	Attribute N
Choice 1	P(1,1)	P(1,2)	...	P(1,j)	...	P(1,N)
Choice 2	P(2,1)	P(2,2)	...	P(2,j)	...	P(2,N)
...	...	...	...	...	...	...
Choice i	P(i,1)	P(i,2)	...	P(i,j)	...	P(i,N)
...	...	...	...	...	...	...
Choice M	P(M,1)	P(M,2)	...	P(M,j)	...	P(M,N)

Fig. 2.4. A typical probability 2−D array, for the PBIL optimization algorithm.

Since each attribute in Fig. 2.4 may be considered as a complete probaobjetcbility event, the sum along each column of the array P must be equal to one, as depicted in Equation (2.3).

$$\sum_{k=1}^{M} P(k,j) = 1, \forall\, 1 \leq j \leq N \qquad (2.3)$$

The main difference between GA and PBIL algorithms, relies on the fact that the PBIL approach stores the search history information on an array like the matrix P of Fig. 2.4, instead of using single and independent representation for each solution of the population. At early iterations of the PBIL algorithm, matrix P shall contain a very disperse set of probabilities, meaning that the associated populations are trying to cover as much as possible of the searching space. This feature is referred to as *exploration* of the solutions space.

Alternatively, when the PBIL algorithm is close to finding a given optimum, probabilities in array P are very concentrated in some of its entries, meaning that those choices are more prone to be selected as the optimal combination. In such final stages of the algorithm, solutions of the generated populations are very concentrated around a region of the searching space, and it is desirable that the searching process converge as fast as possible, for the sake of improving the optimization process speed. Such a feature is referred to as exploitation of the searching space. Exploration and exploitation features are in conflict, since favoring exploration implies a finest and precise search (i.e. solutions of higher quality) but takes more time, whereas exploitation implies faster searches, at expenses of lower quality optimals.

The PBIL algorithm perform the updating of probabilities in array P, by using stochastic information from previous searches which is stored in the array itself, and making a tradeoff between exploration and exploitation features. The key issue in PBIL is to increase those entries of P which seems to be associated to optimal solutions. Such task is achieved by evaluating a different population at each iteration, and looking for the best one in terms of the optimization objectives. When a single entry on the array increases, the remaining entries in the same column must be decreased, in order to cope the constraint stated in Equation (2.3). Algorithm 1 depicts the basic behavior of an adaptive PBIL approach.

Algorithm 1 is consistent with Fig. 2.4, where an instance of a PBIL probability matrix is depicted (called P in such figure). For the sake of allowing full space exploration at the beginning of the algorithm, the whole set of entries are initialized to $1/M$, which means that every choice has the same probability for each attribute, or alternatively, that the whole search space is being considered. If there were any restrictions prohibiting the use of a given choice for a specific attribute, the associated probability in P must set to zero, and the remaining ones in the same column must be initialized in consequence (i.e. to $1/(M-1)$).

After the initialization stage, a loop takes place in the PBIL algorithm, in order to adjust the probabilities in P array gradually. At each iteration (generation) of the algorithm, it is necessary to generate a population of individuals, which is nothing but a set of potential solutions to the optimization problem. The population is generated starting from the probability values in P array, which means that those choices associated to higher probability entries of P are more prone to appear as attributes in the population of individuals. On the other hand, if a given entry on P array has a low probability value, the associated choice shall appear with low frequency in the generated population. A different population is generated at each algorithm iteration, by means of the *Create Population* routine, as shown in Algorithm 1 (line 4). The creation of a different population at each generation, is a major difference of PBIL algorithm with respect to GA, and allows to guaranteeing the population diversity (i.e. it favors the searching space exploration).

Algorithm 1: Adaptive PBIL algorithm.

Input: A probability matrix P, of dimensions $M \times N$, and a Tolerance value for stopping the algorithm
Output: An optimized solution for the problem at hand

1 begin
2    $P(i,j) = \frac{1}{M}; \ \forall \ 1 \leq i \leq M$ and $1 \leq j \leq N$;
3   repeat
4      $Pop = Create\_Population\,(P)$;
5      $Fitness = Evaluate\_Population\,(Pop)$;
6      $Best = Choose\_Best\,(Pop, Fitness)$;
7      $E = Entropy\,(P)$;
8      $LR = Learning\_Rule\,(E)$;
9      $P = Update\_Array\,(P, Best, LR)$;
10   until $(E > Tolerance)$;
11   return $Best$;
12 end

Each individual of the recently created population, must be ranked in terms of its suitability for solving the optimization problem. An assessment value, which often is referred to as the fitness, must be generated for each potential solution in the population. Such a task is performed by the *Evaluate Population* routine (line 5 in Algorithm 1). Several alternatives are available for calculating the fitness values in multiobjective optimization problems. The more common are related to aggregation (normalized and weighted sum) of the objectives and with stochastic approaches.

The *Choose Best* routine, which appears in line 6 of Algorithm 1, looks for the best ranked solution in the population, by using as sorting criterion the individual's fitness values. Such a so called *Best* solution in the algorithm, will be used to update the probabilities in P array.

Two stages are necessary prior to the probabilities updating, for the sake of dynamically adjusting the algorithm's convergence speed. First of all, it is necessary to compute somehow a convergence status of the optimization process. These computations shall serve to favor searching space exploration or exploitation, depending on the current algorithm status. Such status computation is performed by the *Entropy* routine in line 7 of Algorithm 1. As already mentioned, at early iterations of the algorithm, probabilities in the array are very disperse, and at the end of the convergence process, the probability values trend to be very concentrated in some entries of array *P*.

Entropy definition from the information theory serves to assess algorithm convergence status, by treating each column of array *P* as a single random event. According to Claude Shannon's definition [33], Entropy computation for a single column of matrix *P*, would take the form depicted in Equation (2.4). In such equation, let us assume that the Entropy value calculation will performed for a given column *k*.

$$E_k = -\sum_{i=1}^{M} P(i,k) \cdot Log_M[P(i,k)]$$

(2.4)

The total Entropy (*E*) for array *P* may be calculated as the average of the whole columns' Entropy (*E_k*), as shown in Equation (2.5).

$$E = \frac{1}{N} \cdot \sum_{k=1}^{N} E_k$$

(2.5)

Given the described conditions, the value of Entropy (E) will be equal to one when the probabilities in P array are perfectly distributed. This situation occurs at the beginning of the PBIL algorithm, when in the initialization stage all the entries of array P are set to $1/M$ (line 2 in Algorithm 1). In the same way, value of E shall trend to zero as the probabilities in the P array become more concentrated on single entries.

The formulation provided by Equation (2.4) for Entropy computation is not unique. There is a more general formulation referred to as Renyi's Entropy, which may take several forms depending on the value of order used [34]. Renyi's Entropy is a generalization of Shannon, Hartley, min, and collision entropies. Given a single column k of array P (or any other random discrete event), and the conditions described for Equation (2.4), the associated Renyi Entropy may be computed as:

$$E_k(\alpha) = \frac{1}{1-\alpha} \cdot Log_M \left[\sum_{i=1}^{M} P(i,k)^{\alpha} \right]$$

$$(2.6)$$

In Equation (2.6), $E_k(\alpha)$ represents the Renyi Entropy calculated along the column k of array P. α corresponds to the so called entropy order, and $P(i,k)$ for $i = 1,2,...,M$, corresponds to the entries of matrix P along column k.

The order α may be used for calculating several types of Entropy. In fact, when $\alpha = 0$, the formulation of Equation (2.6) is associated with Hartley Entropy. When $\alpha \rightarrow 1$, the formulation of Equation (2.6) takes the form of Shannon entropy. Collision entropy may be obtained for $\alpha = 2$ (this case is often referred simply to as the Renyi entropy), and min-entropy may be derived when $\alpha \rightarrow \infty$. Fig. 2.5 depicts the Renyi entropy for a two-valued random variable (where each potential value has a probability of p and $1-p$, respectively), and several values of α.

As shown in Fig. 2.5 exemplified for $M = 2$, higher orders have the effect of sharpening the entropy behavior, around its maximum value. Alternatively, it can be said that entropy decreases more quickly from its maximum value as the order (α) rises.

Apart from the Entropy computation, the adaptive feature of the PBIL algorithm demands that depending on the convergence status of the searching process, either the exploration or exploitation features to be favored. In other words, by taking into account the entropy values (which measures how close is the searching process of reaching convergence), the algorithm must set a tradeoff between convergence speed or quality. This is performed by means of the Learning Rate parameter or LR. Equation (2.7) shows the probability updating process, which is performed by the *Update Array* routine in Algorithm 1, and is a modified version of the Hebbian Rule [35]. Such a routine must increase the probabilities associated with the best solution found in the population, and decrease the remaining entries in the same column, in order to cope with the constraint established in Equation (2.3).

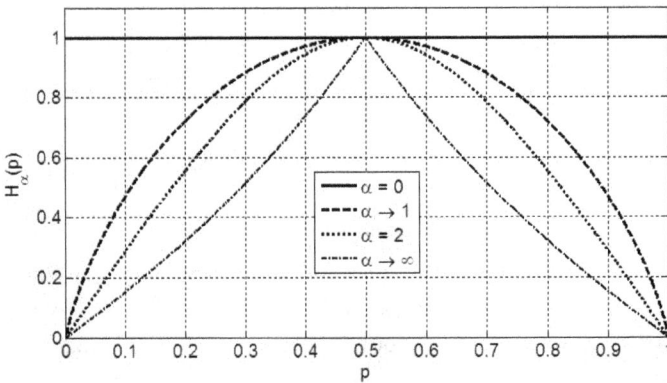

Fig. 2.5. Behavior of the Renyi entropy, for several values of α.

$$P_{(i,k)New} = \begin{cases} P_{(i,k)Old} + \left(1 - P_{(i,k)Old}\right) \cdot LR & if \; i = i_b \\ \left(1 - P_{(i_b,k)New}\right) \cdot \dfrac{P_{(i,k)Old}}{1 - P_{(i_b,k)Old}} & if \; i \neq i_b \end{cases} \tag{2.7}$$

In Equation (2.7), it is supposed that for a given attribute k (k column), the best solution obtained is the choice i_b (row i_b). Suffixes *Old* and *New* in Equation (2.7) are meant to denote the old and new versions of each probability, respectively. As observed in the equation, the amount of increasing performed to the $P(i_b,k)$ entry is proportional to the *LR* parameter. Higher values of *LR* will increase probabilities very quickly, at expenses of coarse searching (search space *exploitation*), whereas lower values of *LR* will perform a finest searching, which entails longer searching times (search space *exploration*).

For the adaptive feature of the PBIL algorithm, the way in which the velocity of convergence (*LR* parameter) is adjusted as a function of the status of convergence (*Entropy* value), is referred to as the *Learning Rule*. Learning Rule computations take place in the *Learning Rule* routine, at line 8 in Algorithm 1. Several learning rules have been proposed in literature, though those which set low values of *LR* at the beginning of the searching process, and speed up the convergence at the end with higher values of *LR*, seems to exhibit best results [36]. Three instances of such ever increasing Learning Rules (namely Linear, Exponential, and Sigmoidal) are presented in Fig. 2.6, along with conitionbell-shaped alternative.

In Fig. 2.6, the LR parameter is meant to vary between LR_{Min} and LR_{Max}, and the Entropy axis (abscissa) has been rearranged in order to reflect what happens as the PBIL searching space operates: Entropy is equal to one at the beginning of the algorithm, and converges to zero at final stages of the searching process. Lower values of *LR* are set at the beginning of the searching, for the four Learning Rules, which logically favors the exploration feature (i.e. fine searching, lower speed). As the convergence progresses, *LR* parameter is increased in order to speed up the search, favoring in this way exploitation of the searching space (higher speed). Bell-shaped rule enhances exploration again at the end of the searching process, which does not seem to be a good strategy, according with reported results [36].

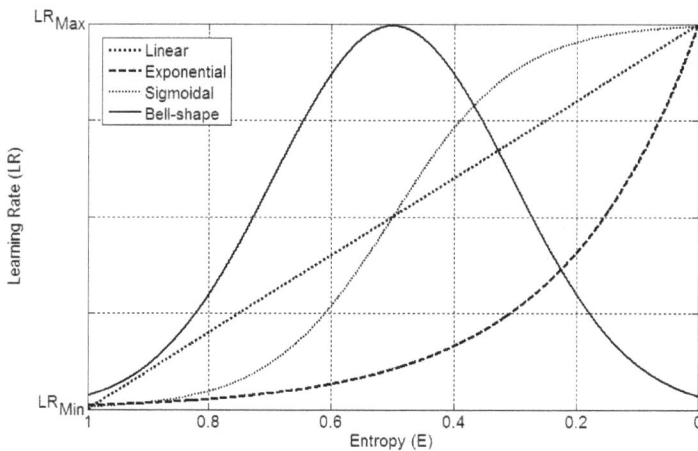

Fig. 2.6. Several instances of Learning Rules for the Adaptive PBIL algorithm.

The iterative process just described and depicted in Algorithm 1, repeats until a suitable value of entropy is reached. Since waiting until entropy value to be exactly equal to zero is very restrictive, the PBIL stopping condition generally takes of an inequality, which compares the entropy value with respect to a given tolerance. The optimization result may be taken of the best solution found at the last iteration, or derived simply from the probabilities array (recall that at this point of the searching process, probabilities in array P are highly concentrated, so deriving the optimal solution just implies finding the maximum values for each column in the array).

2.3. Experimental Results

The PBIL algorithm just described was implemented by using the probability array depicted in Fig. 2.4, and is consistent with Equations (2.1) and (2.2). The later means that the PBIL approach was implemented to solve the task mapping problem of input applications with N tasks, over a target NoC architecture with M nodes, or Processing Elements. Three optimization objectives were taken into account: The completion time (i.e. the execution time of the mapped application running over the target architecture), the power consumption, and the number of hops performed over the network (i.e. the amount of message transactions between neighbour nodes for the transmission of application data), which is a measure of the network traffic. As explained before, such optimization objectives were combined on a single *fitness* value, by means of an aggregation strategy. Shannon Entropy formulation (as shown in Equation (2.4)) was used for computing the convergence state of the algorithm.

Fig. 2.7 shows the evolution of the described optimization objectives as a function of the number of iterations of the algorithm. The input application was obtained from a synthetic annotated ADAG generator, referred to as TGFF [37]. The input annotated ADAG, which is similar in nature to the one depicted in Fig. 2.3, had a mean size of thirty tasks, which are meant to be mapped to a 2D mesh NoC, with a size of 5 by 5 nodes. After a tuning process, the learning rate parameter was set to vary between 0.15 and 0.4, by means of a Linear Learning Rule (as depicted in Fig. 2.6).

In Fig. 2.7, completion time is given in seconds, whereas power consumption is given in watts. As depicted in Fig. 2.7, the three optimization objectives approach to their minimum values, as the time evolves, despite there are some temporary increasing, as a result of the exploration feature. Such increasing and falling to lower values may be viewed as the algorithm escaping from a local minimum.

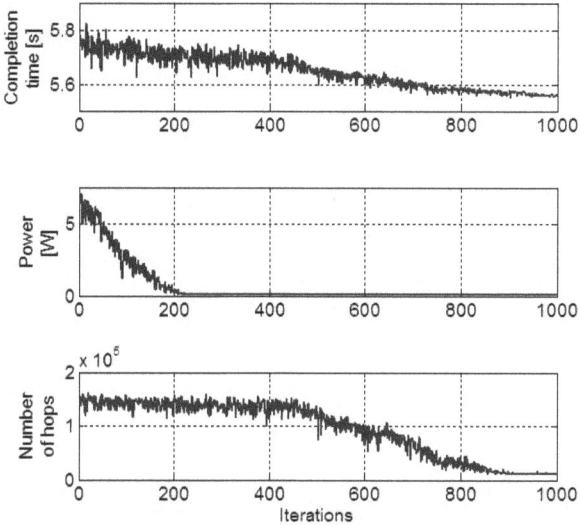

Fig. 2.7. Evolution of the optimization objectives as the algorithm converges.

A valid concern related to the results reported in Fig. 2.7, has to do with the quality of the solutions found by the PBIL algorithm. The values reached after the algorithm convergence are indeed minimum when compared to the history or evolution of the objectives, but a doubt may arise regarding if it does not exist a better combination of parameters, with lower fitness values. Table 2.1 shows the comparison results of the PBIL algorithm, with respect to a Multiobjective Evolutionary Algorithm (MOEA), for a problem of tasks partitioning (which is a problem very similar in nature to tasks mapping) [14]. MOEA is a generalization of Genetic Algorithms (GA), and is a well-known optimization tool, which may be used as a reference.

In Table 2.1, the aggregation function used as fitness for the optimization process takes into account four objectives, namely: Chip area, power consumption, implementation costs, and execution time. After several runs of both algorithms, the four best fitness values were reported in

Table 2.1, and the best for each optimization alternative is underlined, for comparison purposes. As a conclusion, it may be said that there is not remarkable difference between the quality of solutions obtained with the MOEA approach, with respect to those solutions found by the PBIL algorithm. In fact, the four partitioning solutions found by the PBIL algorithm have lower fitness values than those found by the MOEA. It also must be said that convergence time for the PBIL algorithm was on the average around sixty percent of the MOEA convergence times, which means that good enough solutions may be found by the PBIL approach, without the time overhead related to the MOEA approach.

Similar results to those depicted in Table 2.1 have been reported in literature, when comparing the PBIL algorithm with respect to deterministic optimization approaches such as Mixed–Integer programming (MIP) for task mapping problems in NoC architectures [38]. Moreover, the MIP performance degrades very fast when dealing with bigger size problems, just because such approach explores the searching space exhaustively, for the sake of finding the absolute minimum of the fitness.

Table 2.1. A comparison between Multiobjective Evolutionary Algorithm (MOEA) and PBIL algorithm [14].

Optimization Algorithm	Aggregation Function
Multiobjective	337.99
Evolutionary	344.11
Algorithm	324.86
(*MOEA*)	315.96
Population–	311.55
Based	314.21
Incremental	315.44
Learning (*PBIL*)	314.07

Finally, with respect to the adaptive behavior of the PBIL algorithm, and the learning rules depicted in Fig. 2.6, Fig. 2.8 shows the mean convergence times for several mapping problems, as a function of the optimization problem size (i.e. the amount of tasks to be mapped). As a matter of comparison, the four learning rules depicted in Fig. 2.6 (namely: linear, sigmoidal, exponential, and bell-shaped) were profiled in the adaptive version of the PBIL approach, as shown in Algorithm 1.

A MOEA implementation was also profiled, for the sake of providing a reference.

As shown in Fig. 2.8, MOEA and adaptive PBIL for bell-shaped learning rule, exhibit prohibitive convergence times, as the problem size grows. Regarding MOEA implementations, the need of keeping an independent representation for each individual of the population, and to perform evolutionary operations for those representations, may be the reason why there are several orders of magnitude of difference with respect to the performance of adaptive PBIL alternatives. Such alternatives use a very compact representation for the population of individuals, which take the form of a probability array (as the one depicted in Fig. 2.4), and stores all the information necessary for performing an efficient searching space exploration.

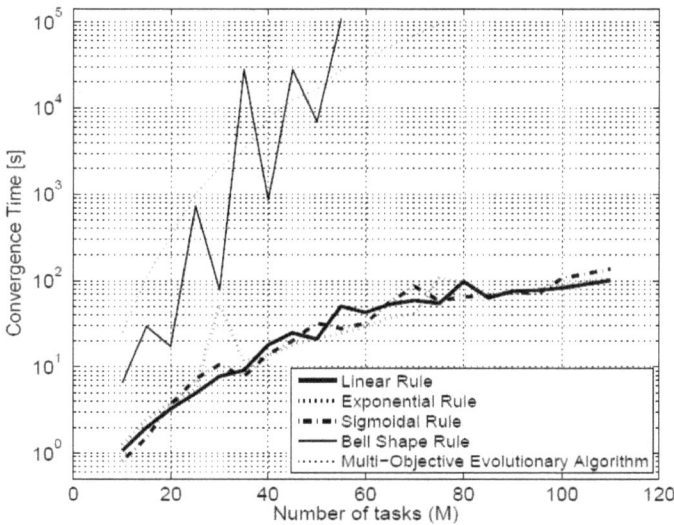

Fig. 2.8. Convergence times for several optimization alternatives.

Regarding the bell-shaped learning rule, the fact of performing space exploration at final stages of the algorithm may be the reason why the corresponding convergence times are very poor when compared to the alternatives with ever increasing learning rates (i.e. linear, exponential, and sigmoidal). Exponential learning rule seems to be less predictable than the linear, and sigmoidal one (notice the convergence time peak of the exponential curve at around a size problem of thirty tasks). In fact, some reported results suggest that there is a foreseeable behavior in the PBIL optimization convergence time, with respect to the problem size (given either as the amount of application tasks, or as the number of PEs present in the target NoC) [32].

2.4. Concluding Remarks

We have shown the feasibility of the Population-Based Incremental Learning (PBIL) Algorithm, for performing optimal task mapping over a target NoC architecture. The experimental results points toward the fact that good solutions may be found by means of the PBIL approach, without incurring in excessive convergence times, as those found for MOEA and MIP optimization approaches.

Both the fact of using a population of solutions for exploring several regions of the searching space simultaneously, and the compact representation for the information collected by such searching process (by means of probabilities), seem to be the reason why PBIL exhibit a good tradeoff between convergence time and quality of the found solutions. The adaptive behavior, may be controlled by means of setting up the limits of the LR parameter, and the learning rule which stablish how to vary such parameter.

The tuning of the adaptive PBIL approach deals only with a few decisions, namely: population size, limits for the LR parameter, and learning rule. This feature makes such approach very suitable for several optimization problems.

Linear and sigmoidal learning rules seem to exhibit the best and more predictable performance. Foreseeable convergence times may be a very attractive feature, since the allow to compute a mapping time budget, and even deciding if some mapping decisions may be performed in run time, instead of design time.

Acknowledgements

The authors would like to thank to *Universidad Nacional de Colombia* and to *Universidad de Antioquia*, for their support in the development of this work.

References

[1]. Robert R. Schaller, Moore's law: Past, present, and future, *IEEE Spectr.*, 34, 6, June 1997, pp. 52–59.

[2]. The International Technology Roadmap for Semiconductors: 1999 Edition. Semiconductor Industry Association.

[3]. M. Kumar, K. Kumar, S. K. Gupta, and Y. Kumar, FPGA based design of area efficient router architecture for network on chip (NoC), in *Proceedings of the International Conference on Computing, Communication and Automation (ICCCA)*, April 2016, pp. 1600–1605.

[4]. S. Bansal, S. Sharma, and N. Sharma, Design of configurable power efficient 2-dimensional crossbar switch for network-on-chip (NoC), in *Proceedings of the IEEE International Conference on Recent Trends in Electronics, Information Communication Technology (RTEICT)*, May 2016, pp. 1514–1517.

[5]. A. Naik and T. K. Ramesh, Efficient network on chip (NoC) using heterogeneous circuit switched routers, in *Proceedings of the International Conference on VLSI Systems, Architectures, Technology and Applications (VLSISATA)*, January 2016, pp. 1–6.

[6]. S. Das, J. R. Doppa, P. P. Pande, and K. Chakrabarty, Energy-efficient and reliable 3D network-on-chip (NoC) Architectures and optimization algorithms, in *Proceedings of the IEEE/ACM International Conference on Computer-Aided Design (ICCAD)*, November 2016, pp. 1–6.

[7]. F. Lokananta, S. W. Lee, M. S. Ng, Z. N. Lim, and C. M. Tang, UTAR NoC: Adaptive network on chip architecture platform, in *Proceedings of the 3rd International Conference on New Media (CONMEDIA)*, Nov 2015, pp. 1–8.

[8]. H. Bokhari, H. Javaid, M. Shafique, J. Henkel, and S. Parameswaran, Malleable NoC: Dark silicon inspired adaptable network-on-chip, in *Proceedings of the Design, Automation Test in Europe Conference Exhibition (DATE)*, March 2015, pp. 1245–1248.

[9]. A. Q. Ansari, M. R. Ansari, and M. A. Khan, Performance evaluation of various parameters of network-on-chip (NoC) for different topologies, in *Proceedings of the Annual IEEE India Conference (INDICON)*, December 2015, pp. 1–4.

[10]. K. Duraisamy, Y. Xue, P. Bogdan, and P. P. Pande, Multicast-aware high-performance wireless network-on-chip architectures, *IEEE Transactions on Very Large Scale Integration (VLSI) Systems*, 25, 3, March 2017, pp. 1126–1139.

[11]. Y. Chen, X. Ling, and J. Hu. A dynamic and low latency wireless NoC architecture, in *Proceedings of the IEEE 11th International Conference on ASIC (ASICON)*, November 2015, pp. 1–3.

[12]. A. Kumar and H. K. Kapoor, Modelling and analysis of wireless communication over networks-on-chip, in *Proceedings of the 18th International Symposium on VLSI Design and Test*, July 2014, pp. 1–6.

[13]. Paris alexandros Mesidis, Mapping of real-time applications on network-on-chip based MPSOCS, Master's Thesis, *University of York*, Heslington, United Kingdom, 2011.

[14]. F. Bolanos, F. Rivera, and J. E. Aedo, Mapping Techniques for Embedded Systems Design with Reliability Considerations, PhD Thesis, *Universidad de Antioquia,* Medellin, 2013.

[15]. Sergiu Carpov, Task mapping and communication routing model for minimizing power consumption in multi-cores, in *Proceedings of the Proceedings of the 8th International Workshop on Network on Chip Architectures (NoCArc '15),* New York, NY, USA, 2015, pp. 27–32.

[16]. Bin Xie, Tianzhou Chen, Wei Hu, Xingsheng Tang, and Dazhou Wang, An energy-aware online task mapping algorithm in NoC-based system, *The Journal of Supercomputing*, 64, 3, 2013, pp. 1021–1037.

[17]. Antunes, M. Soares, A. Aguiar, F. S. Johann, M. Sartori, F. Hessel, and C. Marcon, Partitioning and dynamic mapping evaluation for energy consumption minimization on NoC-based MPSoC, in *Proceedings of the 13th International Symposium on Quality Electronic Design (ISQED)*, March 2012, pp. 451–457.

[18]. Mohammad-Hashem Haghbayan, Anil Kanduri, Amir-Mohammad Rahmani, Pasi Liljeberg, Axel Jantsch, and Hannu Tenhunen, MapPro: Proactive runtime mapping for dynamic workloads by quantifying ripple effect of applications on networks-on-chip, in *Proceedings of the Proceedings of the 9th International Symposium on Networks-on-Chip (NOCS '15),* New York, NY, USA, 2015, pp. 26:1–26:8.

[19]. Changlin Chen and Sorin Cotofana, Link bandwidth aware backtracking based dynamic task mapping in NoC based MPSoCs, in *Proceedings of the International Workshop on Network on Chip Architectures (NoCArc'14),* New York, NY, USA, 2014, pp. 5–10.

[20]. Y. Z. Tei, M. N. Marsono, N. Shaikh-Husin, and Y. W. Hau, Network partitioning and GA heuristic crossover for NoC application mapping, in *Proceedings of the IEEE International Symposium on Circuits and Systems (ISCAS2013)*, May 2013, pp. 1228–1231.

[21]. L. Zhou, M. Jing, L. Zhong, Z. Yu, and X. Zeng, Task-binding based branch-and-bound algorithm for NoC mapping, in *Proceedings of the IEEE International Symposium on Circuits and Systems*, May 2012, pp. 648–651.

[22]. Ou He, Sheqin Dong, Wooyoung Jang, Jinian Bian, and David Z. Pan, UNISM: Unified scheduling and mapping for general networks on chip, *IEEE Trans. VLSI Syst.*, 20, 2012, pp. 1496–1509.

[23]. Amin Rezaei, Masoud Daneshtalab, Maurizio Palesi, and Danella Zhao, Efficient congestion-aware scheme for wireless on-chip networks, in *Proceedings of the 24th Euromicro International Conference on Parallel, Distributed, and Network-Based Processing (PDP)*, 2016, pp. 742–749.

[24]. A. Rezaei, M. Daneshtalab, D. Zhao, F. Safaei, X. Wang, and M. Ebrahimi, Dynamic application mapping algorithm for wireless network-on-chip, in

Proceedings of the 23rd Euromicro International Conference on Parallel, Distributed, and Network-Based Processing, March 2015, pp. 421–424.

[25]. Luciano Ost, Marcelo Mandelli, Gabriel Marchesan Almeida, Leandro Moller, Leandro Soares Indrusiak, Gilles Sassatelli, Pascal Benoit, Manfred Glesner, Michel Robert, and Fernando Moraes, Power-aware dynamic mapping heuristics for NoC-based MPSoCs using a unified model-based approach, *ACM Trans. Embed. Comput. Syst.*, 12, 3, April 2013, pp. 75:1– 75:22.

[26]. Hung-Lin Chao, Sheng-Ya Tung, and Pao-Ann Hsiung, Dynamic task mapping with congestion speculation for reconfigurable network-on-chip, *ACM Trans. Reconfigurable Technol. Syst.*, 10, 1, September 2016, pp. 3:1–3:25.

[27]. Tahir Maqsood, Sabeen Ali, Saif U.R. Malik, and Sajjad A. Madani, Dynamic task mapping for network-on-chip based systems, *J. Syst. Archit.*, 61, 7, August 2015, pp. 293–306.

[28]. E. Khajekarimi and M. R. Hashemi, Energy-aware ILP formulation for application mapping on NoC based MPSoCs, in *Proceedings of the 21st Iranian Conference on Electrical Engineering (ICEE)*, May 2013, pp. 1–5.

[29]. M. Sacanamboy-Franco, F. Bolanos-Martinez, A. Bernal-Norena and R. Nieto-Londono, Genetic algorithm for task mapping in embedded systems on a hierarchical architecture based on wireless network on chip (WiNoC). *DYNA*, 84, 201, 2017, pp. 202–209.

[30]. Adrian Racu and Leandro Soares Indrusiak, Using genetic algorithms to map hard real-time on NoC-based systems, in *Proceedings of the 7th IEEE International Workshop on Reconfigurable Communication-centric Systems-on-Chip (ReCoSoC)*, 2012, pp. 1–8.

[31]. Shummet Baluja, Population-based incremental learning: A method for integrating genetic search based function optimization and competitive learning, *Technical Report*, Pittsburgh, PA, USA, 1994.

[32]. Freddy Bolanos, Jose Edison Aedo, Fredy Rivera, and Nader Bagherzadeh, Mapping and scheduling in heterogeneous NoC through population-based incremental learning, *J. UCS*, 18, 2012, pp. 901–916.

[33]. Lihua Wang, Liangli Ma, Qiang Bian, and Xiliang Zhao, Rough set attributes reduction based on adaptive PBIL algorithm, in *Proceedings of the IEEE International Conference on Information Theory and Information Security*, Dec 2010, pp. 21–24.

[34]. J. Acharya, A. Orlitsky, A. T. Suresh, and H. Tyagi, Estimating renyi entropy of discrete distributions, *IEEE Transactions on Information Theory*, 63, 1, January 2017, pp. 38–56.

[35]. Ray H. White, Competitive Hebbian Learning 2: an Introduction, in *Proceedings of the Proceedings WCNN'93 - Portland*, 1993, pp. 557–571.

[36]. Freddy Bolanos, Jose Edison Aedo, and Fredy Rivera, Comparison of learning rules for adaptive population-based incremental learning

algorithms, in *Proceedings of the International Conference on Artificial Intelligence (ICAI'12)*, 2012, pp. 244–251.

[37]. Robert P. Dick, David L. Rhodes and Wayne Wolf, TGFF: Task graphs for free, in *Proceedings of the 6th IEEE International Workshop on Hardware/Software Codesign (CODES/CASHE '98)*, Washington, DC, USA, 1998, pp. 97–101.

[38]. M. Sacanamboy, L. Quesada, F. Bolanos, A. Bernal, and B. O'Sullivan, A comparison between two optimisation alternatives for mapping in wireless network on chip, in *Proceedings of the IEEE 28th International Conference on Tools with Artificial Intelligence (ICTAI)*, November 2016, pp. 938–945.

3.

From Static to Dynamic: A New Methodology for Development of Simulation Applications

Alexey Cheptsov

3.1. Introduction

The organization of modern Industrial Technical Objects (ITO) is increasingly evolving towards the perspective offered by the Cyber-Physical Systems (CPS) concept [1]. CPS differs from the standard approaches by offering a technology stack that allows integration and collaborative functioning of the physical components (like sensors, producing data, or automatic control systems, consuming those data) with the digital (IT) systems via standardized communication interfaces. One of the goals of such integration is improvement of the decision making process by enabling more complex algorithms (implemented as software services, often open-source) as compared with the ones provided by conventional control systems (developed as proprietary Application Specific Integration Circuits – ASIC firmware) with the per-design limited functionality [2].

The much broader communication and processing capabilities of CPS allow a better integration of different information flows representing the technological processes, also in real time, and thus provide a powerful platform for applications that aim to optimize those processes as well as their numerous non-functional properties, e.g. the power consumption. The CPS concept has found a wide take-up by the industry for building so called "factories of future", as followed for example by a major German innovation program "Industry-4.0" [3].

Alexey Cheptsov
High Performance Computing, Center Stuttgart, Germany

Existing ITO are also taking advantages of the CPS-based organization approach by adopting powerful (in terms of performance) and efficient (in terms of energy consumption) IT platforms and services into their native control procedures and information flows, in order to achieve substantial improvement over the static (pre-defined) ITO configurations (e.g. in case of the unexpected emergency situations, which are not covered by the static "emergency response plans").

Mathematical modeling and computer simulation are the most important scientific methods of projecting and optimization. Many ITO rely on simulation technologies and actively use them in the design phase. On the contrary, there is no or only little use of simulation during the longest phase of the ITO life-cycle (the production). For this reason, the major simulation tools are of a static nature – the models are tailored to the specific ITO prior to their use and their parameters do not change once defined by the experts at the beginning of the analysis. The property of dynamicity, as promoted by the CPS concept, meaning the possibility to obtain the ITO-specific data continuously in a "live-stream" [4] fashion and use them to improve the quality of the technological processes in real-time, offers a fundamentally new perspective for the use of simulation technology – namely to support the ITO that are already in the production state.

The simulation support aims at prediction of possible development of the physical processes based on i) the information on their current state (can be obtained from the sensors) and ii) control setting that are planned to be applied (can be obtained from the control system), leveraging the complex mathematic models developed by the experts. Such a support can be useful in various scenarios. One of them is remedial actions planning in case of unexpected emergency situations. Control systems generally tend to have a very limited "prediction window" for the future situation prediction, which is offset by training activities like machine learning with neuronal networks or fault-based identification [5-6]. Simulation can extend the prediction time horizon of the native control algorithms and alert them about any potential dangerous configurations or states that might be reached by the controlled technological processes as a result of applying a specific control setting (see an example of prediction of a safety-critical technological process in Fig. 3.1). The CPS concept allows the simulation to be executed proximately in the controlled environment, thus avoiding the "offline" coupling of the data with the simulation platform in order to facilitate the meeting of real-time processing requirements of the control systems (Fig. 3.2).

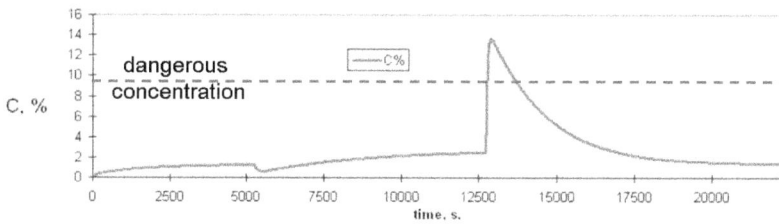

Fig. 3.1. Example of gas concentration (*C*) simulation in a safety-critical ventilation object. The simulation aims to detect over-limit states of the marsh gas concentration.

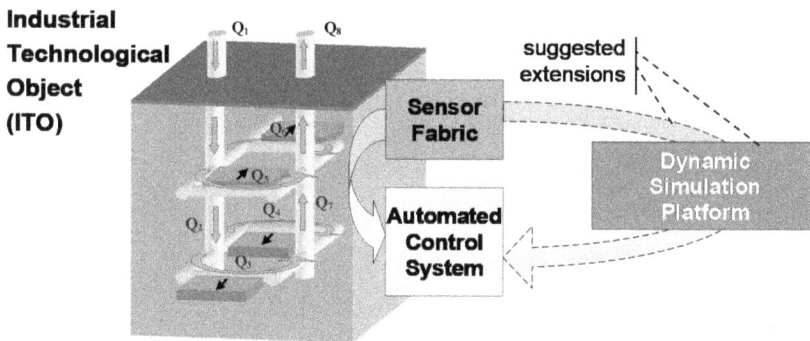

Fig. 3.2. Example of gas concentration (*C*) simulation in a safety-critical ventilation object: The simulation aims to detect over-limit states of the marsh gas concentration.

The current innovative research trends like the Digital Twin [7], Integrated Simulation [8], or Hardware-in-the-Loop [9] although follow the concepts that break through the static nature of the conventional simulation but generally lack the methodology of dealing with complex (e.g. in terms of hierarchical levels of technological processes and their control systems) objects, which requires distributed, component-based, and interoperable approaches to the organization of the real-time simulation environment (see a comparison in Table 3.1). Also from the simulation hardware perspective, the generic hardware that is targeted by the conventional tools is not always possible to be applied in a technological environment due to constraints like power consumption or required performance. Therefore the use of specialized devices which ensure a trade-off between the application requirements and the infrastructure capabilities is required (see some examples in Fig. 3.3).

71

The use of dedicated hardware platforms such as Myriad-2 of Intel [10] or massively parallel systems on chip Zynq of Xilinx [11], is thus necessary. However, for most of the existing simulation tools and platforms, the dedicated heterogeneous hardware, especially in the distributed system context, are impractical due to a limited design of the software frameworks used to implement those platforms and tools.

Table 3.1. Comparison of real-time data analysis technologies.

Technology / Property	Digital Twin	Integrated Simulation	HW in the loop	Dynamic Simulation
Real-time data access	+	+	+	+
Support of complex topologies	+	-	-	+
Decentralized data acquisition	+	-	-	+
Portability to low-power and acceleration hardware	-	-	+	+
Scalability	-	-	-	+
Energy-efficiency guarantees	-	-	-	+

The aim of this chapter is to elaborate the concept of Dynamic Simulation and analyse the major requirements towards creating a software platform implementing this concept on a wide range of the hardware devices. As a basic application domain we are considering the ventilation networks of underground coal mines, which impose challenging requirements of security in the real-time context. On the example of this basic domain, we hope to attract more application domains to take the advantages of the dynamic simulation as well as of the software and infrastructure solutions offered by the EU-project PHANTOM[4].

---

[4] PHANTOM (http://www.phantom-project.org/) is a project receives funding under the European Union's Horizon 2020 Research and Innovation Programme under grant agreement No. 688146.

a)[5]

b)[6]

c)[7]

d)[8]

e)

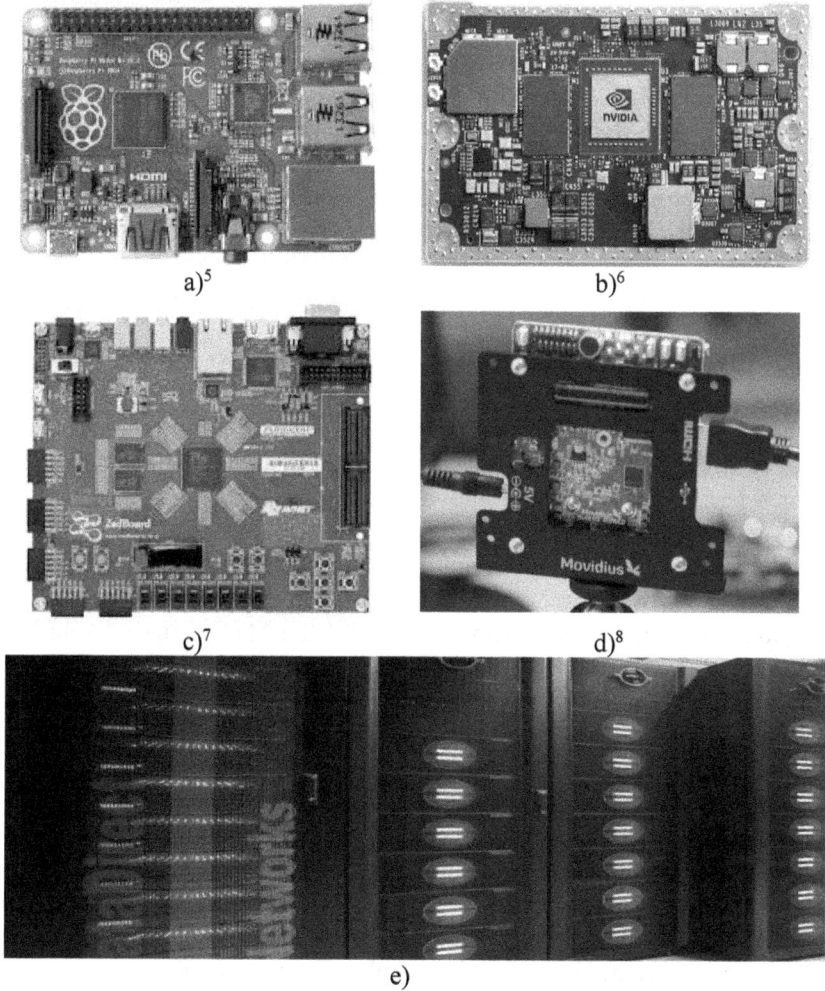

Fig. 3.3. Components of a heterogeneous simulation infrastructure:
a) CPU – ARM-based Raspberry Pi; b) GPU – Jetson-TX2 of NVIDIA;
c) FPGA – Zynq-ZC706 of Xilinx; d) Embedded – Myriad2 Movidius;
e) HPC-infrastructure of HLRS.

---

[5] Source: Raspberrypi.org
[6] Source: https://techcrunch.com/2017/03/07/nvidias-jetson-tx2-makes-ai-computing-possible-within-cameras-sensors-and-more/
[7] Source: Xilinx.com
[8] Source: Digit.in

The remainder of the chapter is organized as follows. Section 3.2 introduces the methodology of dynamic simulation and gives overview of the major requirements of simulation applications to a dynamic simulation platform. Section 3.3 discusses a challenging application domain for the dynamic simulation platform – the underground mine ventilation systems – and shows some representative models for it. Section 3.4 gives an overview of the chosen implementation technology for a dynamic simulation platform prototype. Section 3.5 concludes the chapter.

3.2. Methodology of Dynamic Simulation

The Cyber-Physical Systems (CPS) organization concept promises to endorse real-time data streams of Industrial Technological Objects (ITO) for access not only by the native control systems through proprietary communication channels but also by a wide range of external IT solutions via commodity communication interfaces like Ethernet. The goal of the dynamic simulation is to exploit the potential of the simulation technology to support full-fledged ITO in their production phase, as facilitated by the CPS. The advantages of the dynamic approach as compared to the conventional, static, one are manyfold. First, the precision of the simulation can be greatly improved by the possibility to compare the simulation results with the real data obtained in the real time (e.g. provided by the sensors), which can help optimize the major parameters and coefficients of the models "on the fly", i.e. right at the time of simulation and in a fully automatic fashion. Next, the automatic control systems would get an opportunity to improve their algorithms by incorporating the predictions made by the simulation, which might be performed before applying any essential control setting on the controlled object, which is especially necessary for relatively slow dynamic processes, i.e. with the duration that is long enough to outreach the "prediction window" of the standard control system (like can be found in the majority of gas-dynamics simulations). Another benefit of the dynamic way of performing the simulation is the possibility to support the ITO operators (humans) that take a manual control over the dynamic process in case of emergency situations; in such cases the availability of the real-time simulation support in the decision making loop is essential. Reactive and dynamic simulation requires new approaches to building simulation platforms, which should be modular to support a wide range of diverse physical ITO components, portable to be able to run on the available heterogeneous hardware, distributed to

support complex ITO structures, efficient in terms of performance and
energy consumption of the simulation hardware.

The goal of achieving the dynamic behavior for the simulation
technologies involves 4 major aspects:

- Incorporation of live data into the simulation models;

- Real-timeliness of the actual situation development forecast;

- Software integrity across heterogeneous and distributed hardware
platforms;

- Support of industrial automation tasks.

Incorporation of live data into the simulation models aims to overcome
the static nature of the conventional simulation platforms like
Matlab/Simulink with the major limitation of the input parameters to be
pre-configured, i.e. defined prior to the simulation start. For the real
technological objects, those parameters never remain constant and are
dynamically changing, which might happen even during the simulation.
This negatively affects the quality of the simulation results. On the
contrary, the property of dynamicity should allow the simulation process
to rely on the most actual data about the investigated object, provided by
the sensors attached to the system. For this purpose, the dynamic
simulation platform should expose a front-end through which the data
can contiguously arrive and be used to steer the simulation, which turns
it into a pretty much service-oriented and contiguously-triggered process
which depends on the incoming data events.

The property of dynamicity enables for the simulation a new set of
challenging usage scenarios, among others – the fault-tolerance
improvement of the fully-fledged automatic control systems, their
misbehaviour prevention, ad-hoc creation of 'emergency response
plans'.

Real-timeliness of the actual situation development forecast implies
that the dynamic simulation platform employs the definition of the real
time and aligns the relative timeline of the simulation with the absolute
one of the real-time while representing the simulation results. The notion
of the real time enables the simulation platform to precisely define, for
which time range (starting from the current time) the simulation results

have their validity. This allows the dynamic simulation platform to organize the simulation experiments i) Continuously and ii) Contiguously in order to ensure that the forecast for the user-defined "prediction window", lying in the future, are always in the most actual state. A special simulation experiment planner should be provided by the dynamic simulation platform for this purpose.

Software integrity across heterogeneous and distributed hardware platforms is one of the most challenging requirements of the dynamic simulation. The software implementation of the numerical algorithms, as required by most of the mathematical models, needs powerful computer workstations to run on. However, those facilities cannot be integrated into the security-critical ITO-infrastructure with a high demand of vibration level, humidity, dust and gas emissions, and other technological factors preventing the use of the commodity hardware. Also the distributed nature of the ITO sensor network makes the "centralized" simulation impossible. The technological systems ("embedded") hardware is often composed by dedicated systems, which are of a smaller size, consume less energy than the commodity systems, but also reach a much lower performance. Such systems cannot host the computation-intensive simulation applications. However, within the context of the whole ITO, there are many such devices, which make up a distributed system, interconnected by a productional network like CAN bus or by a more generic one ETHERNET (see an example in Fig. 3.4). Their consolidation into a common deployment infrastructure allows achieving the acceptable performance characteristics by the simulation application.

The dynamic simulation software should be able to deploy the simulation services on the distributed hardware devices, taking into account their availability, current utilization, energy consumption limitation policies, etc. Moreover, the distributed nature of the dynamic simulation platform's components allows for a straightforward integration of "external" compute platforms like cloud or high-performance computing systems, which can be used for very complex simulations while keeping the more simple ones on the embedded devices.

Support of industrial automation tasks implies that the control systems can interact with the dynamic simulation platform in order to launch new simulation experiments or request the results of the already completed ones (in the real time). For this, the dynamic simulation platform should provide results in the form that would be easily accessible by microcontrollers. For this, standard interfaces like JSON or XML should be employed, in order to simplify the communication protocol. The control system will thus be offered a possibility to integrate the dynamic simulation platforms into its native control- and data-flow communication chains.

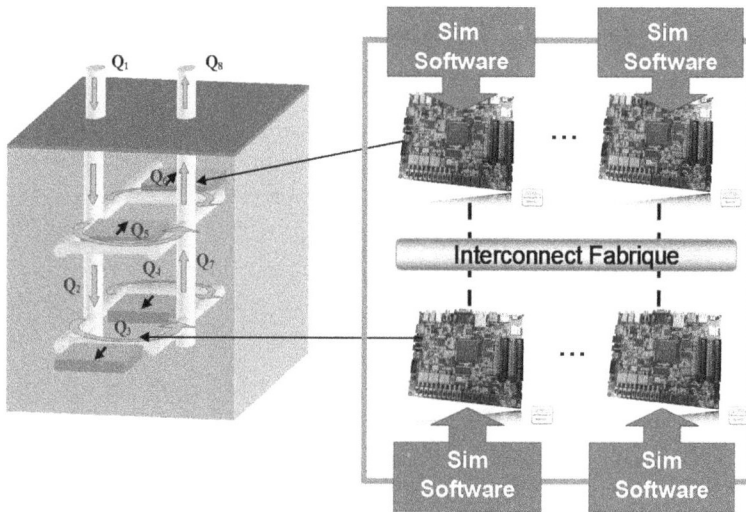

Fig. 3.4. Distributed hardware environment for dynamic simulation.

3.3. Underground Mine Ventilation Systems as Objects of Dynamic Simulation

3.3.1. General Overview

On the search for a challenging application domain for the dynamic simulation platform we could not come past by industrial ventilation systems, which consist of a large number of complex, multidivisional individual objects, each described by high-complexity mathematical models. The coal-mine ventilation was selected due to availability of security-critical processes and, accordingly, of the related complex automatic controlling tasks.

The ventilation in coal mines aims to ensure a fail-safe production in underground mining areas [12]. Notably, the necessary limits of air flow, gas emission (e.g. marsh gas - CH4) concentration, and temperature should be ensured in the production areas (referred as "**Faces**" in the coal-mining terminology [13]). The ventilation is performed by an admission of the fresh air flow from the coal mine surface, by means of special air-fans [14], through the transportation elements – the **Roadway** and the **Gateway**. The Face, the Roadway, and the Gateway form a typical **Ventilation Section** of a mine ventilation system (Fig. 3.5). Due to the availability of a filtration space that verges on the face, roadway, and gateway – the Goaf – the Ventilation Section is also a subject of the gasdynamics.

The ventilation systems of a real-complexity ventilation object can contain hundreds of sections, connected into a common topology, which can be represented by a directed acyclic graph (Fig. 3.6).

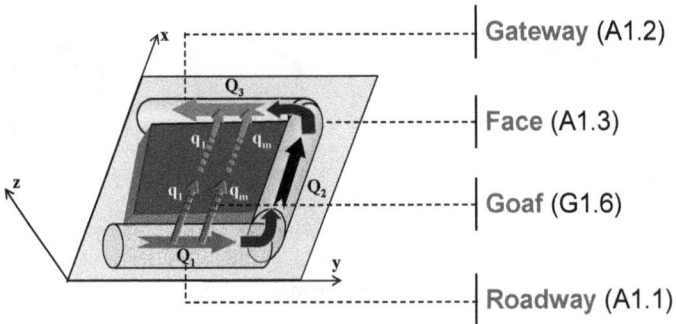

Fig. 3.5. Simplest ventilation section.

The necessary volume of the air flow through all sections of the ventilation network is controlled by an automatic system that is built based on conventional "closed-loop" or "feedback-control" approaches [15-16].

The input parameters for the control are the values of the actual (at any moment of time) air consumption and gas concentration, which are measured by the sensors [17] that are installed locally in the object of measurement (usually – the gateway of the ventilation section). The control settings are applied by the control system either locally by means of the regulators installed in the controlled object, or globally by the ventilators. The local regulation is usually implemented as a vertical pull-out door with a free clearance through which the air is flowing. In case if the local regulation does not ensure achieving the necessary values of the regulation parameters [18], the global regulation setting are applied, fostering the change of the global ventilator(s) rotation speed.

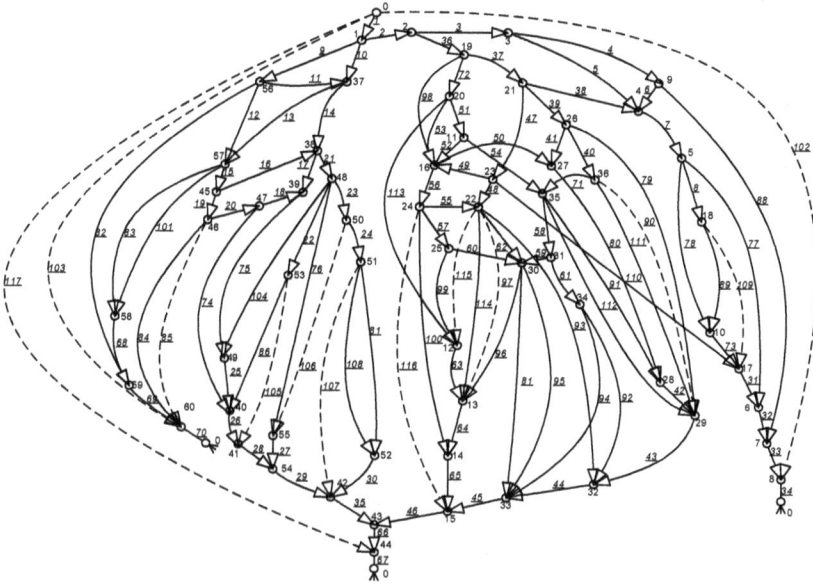

Fig. 3.6. Real-complexity topology of coal-mine South-Donbass-3 (located in Ukraine) ventilation system with m=117 ventilation sections and n=61 connection nodes.

Underground coal mines and in particular their ventilation systems are challenging ITO of the coal industry and, as such, are in focus of many global industrial activities, such as the German "Mining-4.0" program [19], which is coordinating the industrial innovation activities around the coal-mining related problems. The problematic which we are dealing with in this chapter as an exemplary scenario for the dynamic simulation, is also available in the other major global systems with interconnected objects, such as transportation, geology, bioengineering, smart cities and many others.

The dynamic processes of mine ventilation systems can be analysed by means of a quite rich set of mathematical models, which are described in different literature sources. Those models differ in the way treating the space continuity (e.g. systems with concentrated and distributed parameters), the granularity of the analysed "elementary" processes (a ventilation element, a section, or even the whole network as a single model), etc. However, in order to support the full-fledged industrial systems, the models should confirm with the sensor network's information flow structure. Our proposed approach of harmonization of the models for any application domain consists in 3 following actions:

- Classification of the controlled objects for which the models are available;

- Elaboration of model interfaces for each of the objects, based on the information flow within the analysed cyber-physical system;

- Specification of the API to implement the wrap the models into a simulator – a software container for executing the model and organizing all information dependencies.

3.3.2. Exemplary Models

Let's consider a model of a typical Ventilation Section (cf. Fig. 3.5), which is comprised of the following elements: the Roadway, the Face, the Gateway, and the Goaf, as described in Section 3.1. This model describes dynamical change of the air- and gas-flow that results from applying a new regulation set to the Ventilation Section.

The overall (air- and gas-dynamic) model of the Ventilation Section is defined by the parameters in Table 3.2 (with the following types: I - input, O - output).

The aerodynamic processes in the Face, the Roadway, and the Gateway are described by the following basic mathematic model (a special form of the Navier-Stokes Equation):

$$\begin{cases} -\dfrac{\partial P}{\partial \xi} = -\dfrac{2\rho}{F^2} Q \dfrac{\partial Q}{\partial \xi} + \dfrac{\rho}{F}\dfrac{\partial Q}{\partial t} + rQ^2 + r'(t)Q^2 \\ -\dfrac{\partial P}{\partial t} = \dfrac{\rho a^2}{F}\dfrac{\partial Q}{\partial \xi} - \dfrac{\rho a^2}{F} q \end{cases} , \qquad (3.1)$$

where ρ, F, r are the coefficients, the rest are the parameters from Table 3.2.

Table 3.2. Parameters of the Ventilation Section.

Parameter	Type	Description	Unit
$H(\Delta P)$	I	The depression (difference pressures) between the edges of the ventilation section (i.e. between the begin of the roadway and the end of the gateway)	Pa
r'	I	The airdynamic resistance of the local regulation body, usually installed in the gateway	Ns^2/m^8
QoM	I	The debit of the marsh gas at the initial time of the simulation	m^3/s
Q	O	The airflow in the ventilation section (i.e. in the road- and gateways)	m^3/s
Qs	O	The airflow in the face	m^3/s
q	O	The airflow of leaking in the goaf	m^3/s
C	O	The concentration of the marsh gas in the section (i.e. in the road- and gateways)	%
Cs	O	The concentration of the marsh gas in the face	%

Very commonly, an approximation of the model (3.1) is used, as proposed by Svjatnyj in the work [20]. In Svjatnyj's model the compressibility of air as well as filtration flows are disregarded, so that the basis model (3.1) is conversed into the simplified representation (3.2).

$$K\frac{dQ}{dt} + (r + r'(t))Q^2 = H(t)$$
,
(3.2)

where K is the coefficient of the air inertia.

The gasdynamic processes in the Goaf are described by the following mathematical model (3.3):

$$V_S \frac{dC_S}{dt} + (Q_S + Q_{MS}) \cdot C_S = Q_{MS}$$

$$V_{FR} \frac{dQ_M}{dt} + (Q_M - Q_{0M}) \cdot q = \frac{V_{FR} \cdot Q_M}{q} \frac{dq}{dt}$$

$$V_{VS} \frac{dC}{dt} + (Q + (Q_S + Q_{MS}) \cdot C_S + Q_M) \cdot C = (Q_S + Q_{MS}) \cdot C_S + Q_M \quad (3.3)$$

where Q_M is the debit of the marsh gas, V_{FR} – volume of the goaf in which the mix of the air and of the marsh gas happens, V_{VS} – the volume of the gateway in the air-gas mix happens.

The simulation model, resulting from the application of a numerical method (e.g. Eyler), turns into the Equation system (3.4). The results of simulation are shown in Fig. 3.7.

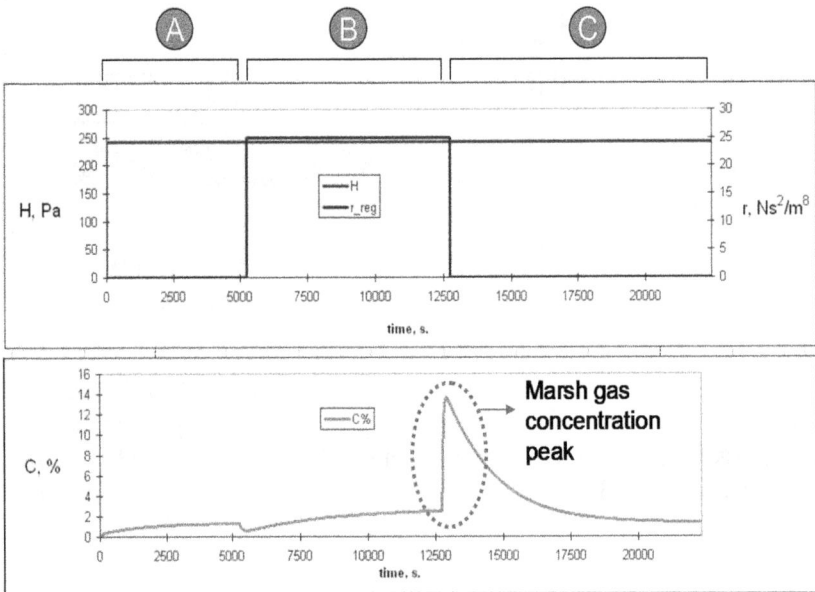

Fig. 3.7. Gas- and Air-dynamic processes in a ventilation section caused by the changing ventilator depression (part A) as well as local regulator settings (parts B and C).

$$
\begin{cases}
Q_{S,t+1} = Q_{S,t} + h \cdot \dfrac{K_q}{K \cdot K_q + K_q \cdot K_S + K \cdot K_S} \\[2ex]
\quad \cdot \left(H - (\dfrac{K+K_q}{K_q}) \cdot r_s \cdot Q_S^2 + \dfrac{K}{K_q} \cdot r_q \cdot q^2 - (r+r') \cdot Q^2 \right) \\[3ex]
q_{t+1} = q_t + h \cdot \dfrac{K_S}{K \cdot K_q + K_q \cdot K_S + K \cdot K_S} \\[2ex]
\quad \cdot \left(H - (\dfrac{K+K_S}{K_S}) \cdot r_q \cdot q^2 + \dfrac{K}{K_S} \cdot r_s \cdot Q_S^2 - (r+r') \cdot Q^2 \right) \\[3ex]
Q_{t+1} = Q_{S,t+1} + q_{t+1} \\[2ex]
C_{S,t+1} = C_{S,t} + h \cdot \dfrac{Q_{MS} \cdot (1-C_S) - Q_S \cdot C_S}{V_S} \\[3ex]
Q_{M,t+1} = Q_{M,t} + h \cdot \left(\dfrac{Q_M}{q} \cdot (H - r_q q^2_t + (r+r') \cdot Q_t^2) \cdot \dfrac{1}{K_q} - \dfrac{(Q_M - Q_{0M}) \cdot q}{V_{FR}} \right) \\[3ex]
C_{t+1} = C_t + h \cdot \left(Q_S \dfrac{C_S - C_S \cdot C}{V_{VS}} + Q_{MS} \dfrac{C_S - C_S \cdot C}{V_{VS}} + Q_M \dfrac{1-C}{V_{VS}} - Q \dfrac{C}{V_{VS}} \right)
\end{cases}
\qquad (3.4)
$$

The gas leakage into the production area (the Face) is explained by changing masses of the marsh gas (emitted from the filtration space – the Goaf) that are caused by the changing speed of the air flow in the road- and gate-ways [21].

The relative-time results are being aligned with the real-time ones and the mid-term forecast is thus produced (see Fig. 3.8).

The computation (3.4) is being repeated every time the input parameters (see Table 3.2) change.

3.4. Implementation with PHANTOM Framework

The software implementation of the (hierarchically organized) models – the simulators – is proposed to perform following a service-oriented approach. Each simulator thus acts as an independent software service (the Agent) that is responsible for the simulation tasks for the physical object that it is associated with. The Agent communicates with the other, "neighboring", Agents (for any data exchange according to the model specification) or with a supervising component (the Controller), via control messages. Fig. 3.9 shows a simulators architecture for the dynamic simulation platform.

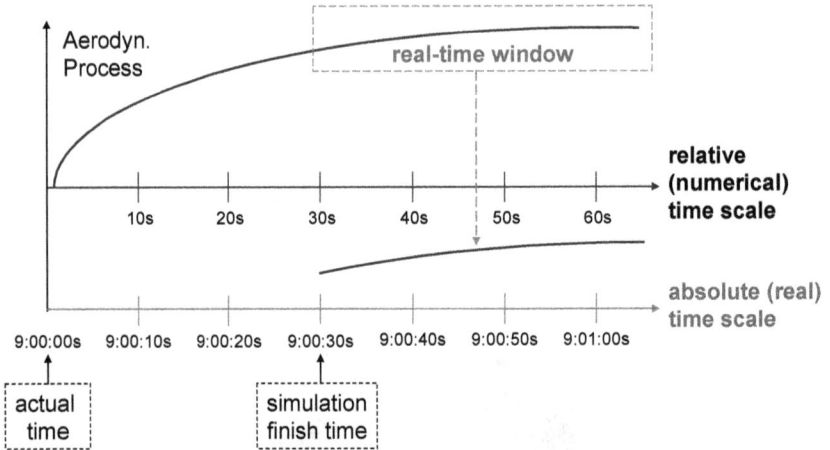

Fig. 3.8. Aligning the simulation results with the real time.

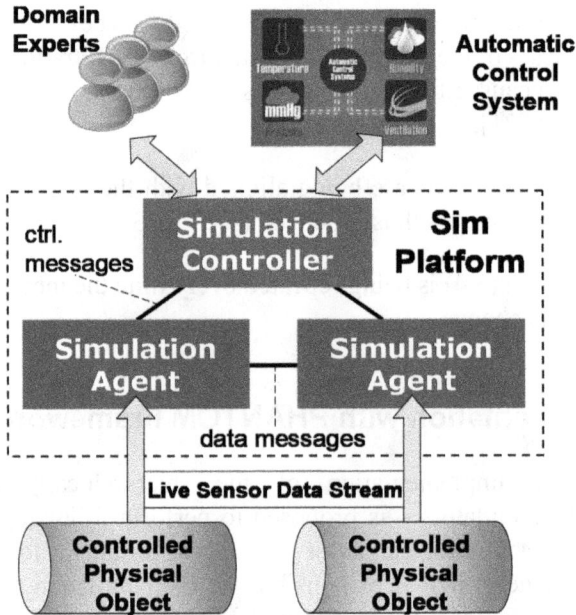

Fig. 3.9. Implementation of simulators as services.

The Agents thus need to expose interfaces (API) for i) incorporating the model's functionality and ii) control- and data-flow communication. The model's functionality is proposed to implement in form of stateless micro-services that receive the input parameters (according to the specification of the model, see the previous Section 3.5) and be triggered by the Agent to perform the simulation cycle whenever instructed by the Controller or in case of the changing input sensor parameters. The Communication Library should support the asynchronous data exchange between the components (the Agents or the Controller).

In view of the heterogeneous nature of the hardware, on which the simulators are supposed to be running, ensuring the interoperability and portability is essential for the implementation of the simulators (regardless whether of the Agents or of the Controllers). For this, a cross-platform compilation and execution environment that is developed by the PHANTOM project offers a solution. PHANTOM targets development across the entire computing continuum, from resource-constrained embedded devices up to powerful multi-core compute clusters. It does this by the definition of a component-based programming model in which user applications are comprised of a set of components that are independent, concurrent, micro-services. PHANTOM's programming model is amenable to describing systems in terms of their control flow and data flow; but also wider system concerns like security and the measurement and analysis of non-functional properties. This is achieved by starting with a programming model and runtime suitable for cyber-physical and embedded systems, in which the mapping and data movement decisions are made offline using multidimensional optimization. This model fits very well to the concept of simulators of the dynamic simulation platform, with the predefined configuration of interconnected services (reflecting the physical structure of the modelled object) and potentially heterogeneous deployment hardware for each of them. This model can then be scaled up to meet the cloud use case. Supporting the programming model is a representation of non-functional requirements and characteristics of the underlying heterogeneous platform structure and expected behaviour (e.g. CPU speeds or bus bandwidths) – this will enable effective mapping of the application to the underlying heterogeneous platform to meet application requirements. Security is an important part of the PHANTOM platform, and is considered to be a wide-ranging, multi-level problem that concerns many aspects of hardware, software, and application design.

3.5. Conclusions

Simulation is a powerful scientific technology to support industrial tasks in many aspects. However, the support of fully-fledged industrial technical systems is mainly out of the simulation scope due to the strong static nature of the available simulation tools. This chapter elaborated a methodology of dynamic simulation, which enables the simulation-aided support of the industrial automation tasks in production by leveraging the concept of the simulation as a service. The combination of simulation services into a rich-functional simulation platform, as enabled by the software implementation technologies that were discussed in the chapter, offers a new perspective of the simulation technology evolution.

The innovation lies in elaborating new ways of organization of the simulation process, which is proposed to be service-oriented, reactive, event-triggered, real-time aware, and with the exposed interfaces that allows the integration into the available heterogeneous IT-infrastructure of the targeted technological environment with the "live data traffic". The dynamic simulation concept was proved for the challenging in terms of the presented security and safety demands domain of the underground coal mine ventilation and indicated a good applicability. An implementation concept for the simulation software was proposed based on the service-oriented and component-based approach, as elaborated by the PHANTOM project.

The chapter aimed to introduce the major technologies that are needed to fulfill the vision of the dynamic simulation. A lot of implementation and integration work lies ahead. In particular, the following actions are to be performed:

- Integration with the automatic control systems. The applications should be elaborated that should define the procedure of interaction between the control system and the simulation platform.

- Security concepts have to be elaborated. There are manifold layers and aspects of security which should be addressed in the simulation platform's prototype.

- Performance and energy consumption of the simulation application shell be evaluated. The available hardware platforms will most probably not be able to ensure a high efficiency of the models executed on them.

Therefore tools that should improve the efficiency by a better parallelization, power mode control and other techniques are to be used.

Once successfully implemented for the targeted coal mine ventilation domain, the dynamic simulation methodology will also be promoted to the further domains and more usage scenarios will be attracted.

Acknowledgements

This publication is a result of the PHANTOM project that has received funding from the European Research Council (ERC) under the European Union's Horizon 2020 research and innovation programme (grant agreement No. 688146).

References

[1]. About Cyber Physical Systems (from University of Illinois, Department of Computer Science): http://cs.illinois.edu/research/themes/cyberphysical (retrieved: 08, 2017).

[2]. T. Hamada, K. Benkrid, K. Nitadori, M. Taiji, A Comparative Study on ASIC, FPGAs, GPUs and General Purpose Processors in the O(N^2) Gravitational N-body Simulation, in *Proceedings of the NASA/ESA Conference on Adaptive Hardware and Systems*, 2009, pp. 447-452.

[3]. About Industry 4.0 (whitepaper): https://www.zvei.org/presse-medien/publikationen/industrie-40-whitepaper-zu-forschungs-und-entwicklungsthemen/ (retrieved: 08, 2017).

[4]. I. Akkaya, Data-Driven Cyber-Physical Systems via Real-Time Stream Analytics and Machine Learning, PhD Thesis, EECS Department, University of California, Berkeley, *Technical Report No. UCB/EECS-2016-159,* October 25, 2016.

[5]. M. Bordasch, C. Brand, P. Göhner, Fault-based identification and inspection of fault developments to enhance availability in industrial automation systems, in *Proceedings of the 20th IEEE International Conference on Emerging Technologies and Factory Automation*, 2015.

[6]. B. Bottcher, J. Badinger, O. Niggemann, Design of industrial automation systems - Formal requirements in the engineering process, in *Proceedings of the 6th IEEE International Conference on Emerging Technologies & Factory Automation*, 2013, pp. 1-4.

[7]. M. Grieves, J. Vickers, Digital Twin: Mitigating Unpredictable, Undesirable Emergent Behavior in Complex Systems, in F.-J. Kahlen, *et al.* (Eds.), Transdisciplinary Perspectives on Complex System, New Findings and Approaches, *Springer*, 2017, pp. 85-113.

[8]. W. Yan, Y. Xue, X. Li, J. Weng, T. Busch, J. Sztipanovits, Integrated simulation and emulation platform for cyber-physical system security experimentation, in *Proceedings of the 1st International Conference on High Confidence Networked Systems, 2012*, pp. 81-88.

[9]. M. Bacic, On hardware-in-the-loop simulation, in *Proceedings of the 44th IEEE Conference on Decision and Control, and the European Control Conference 2005 Seville*, Spain, 2005, pp. 3194-3198.

[10]. About Movidius' Myriad2 embedded system: https://www.movidius.com /solutions/vision-processing-unit (retrieved: 08, 2017).

[11]. About XILINX's massively parallel System-on-Chip solution Zynq-7000: https://www.xilinx.com/products/silicon-devices/soc/zynq-7000.html (retrieved: 08, 2017).

[12]. F. Heise, F. Herbst, Coal-mine ventilation, (in German), Chapter in Kurzer Leitfaden der Bergbaukunde, 1932, pp. 86-117.

[13]. About common underground mine ventilation terminology: https://en.wikipedia.org/wiki/Underground_mine_ventilation (retrieved: 08, 2017).

[14]. About underground mine ventilators: https://de.wikipedia.org/wiki/ Grubenl%C3%BCfter (retrieved: 08, 2017).

[15]. K. J. Aström, R. M. Murray, Feedback Systems. An Introduction for Scientists and Engineers, *Princeton University Press*, 2008.

[16]. About major control approaches: http://www.electronics-tutorials.ws/systems/closed-loop-system.html (retrieved: 08, 2017).

[17]. About measurement methods for underground mine ventilation: https://de.wikipedia.org/wiki/Wettermessung (retrieved: 08, 2017).

[18]. K. Hatzfeld, Handbook of coal mine safety, VEB, Leipzig 2. edition, 1960.

[19]. About Mining-4.0: http://www.bergbau-vier-punkt-null.com/ (retrieved: 08, 2017).

[20]. F. Abramov, L. Feldmann, V. Svjatnyj, Modelling of dynamic processes in underground mine ventilation, *Naukova Dumka*, Kiev, 1981, (in Russian).

[21]. V. Svjatnyj, Simulation of aerodynamic processes and development of control system for underground mine ventilation, *PhD Thesis*, 1985, (in Russian).

4.

Improvement of the Calibration Uncertainty for Class E₁ Weights Using an Adaptive Subdivision Method on an Automatic Mass Comparator

Adriana Vâlcu

4.1. Introduction

In 1889, at the First Conference of Weights and Measures (CGPM), the kilograms prototypes were shared - by chance - for each country. Romania has received the "National kilogram Prototype No. 2" (NPK).

NPK is a solid cylinder having a height equal to its diameter (39 mm) and consists of an alloy of 90 % Platinum and 10 % Iridium (Pt-Ir) having a density of approximately 21500 kg/m^3. Now, it is maintained by the National Institute of Metrology and serves as a reference for the entire dissemination of the mass unit in Romania.

The realization and dissemination of the unit of mass by Mass Laboratory of the Romanian National Institute of Metrology starts from the reference stainless steel standards (a set of three 1 kg mass standards and two sets of disc weights from 500 g to 50 g), which are traceable to the International Prototype Kilogram through the mass of the Romanian Prototype Kilogram No. 2.

Starting from these reference stainless steel standards, submultiples and multiples of the unit are realized to permit the masses of additional bodies to be determined with traceability to the international standard. This takes place with the aid of several weights sets E_1 of suitable grading (in most cases 1, 2, 2, 5) which are determined "in themselves" according

Adriana Vâlcu
Regional Directorate of Legal Metrology, Bucharest, Romania

to proper weighing designs and by using a least squares analysis (with subdivision or multiplication methods).

In the calibration of class E_1 weights, when the highest accuracy is required, the subdivision method is mainly used.

The subdivision weighing design has both advantages and disadvantages:

- Advantages [1]:

a) Minimizes handling (and hence wear) of standards;

b) Produces a set of data providing important statistical information about the measurements and the daily performance of the individual balances;

c) Offers a redundancy of data.

- Disadvantages [1]:

a) Requires a relatively complex algorithm to analyze the data (as compared with other methods, for example Borda);

b) Necessitates placing groups of weights on the balance pans (this can cause problems for instruments with poor eccentricity characteristics, or automatic comparators designed to compare single weights).

To apply the calibration by subdivision method on the automatic comparator, a set of disc weights (which represent reference standards) has been used. These weights constitute both support plates and check standards.

The main objective in the search for better designs was to find a calibration scheme which can be performed considering the following factors: the automatic comparator, the diameter of the disc weights (so that a group of OIML weights can be placed over) and the efficiency of design matrix.

The "adaptive subdivision method", presented in this chapter, allows the cylindrical weights with a lifting knob, having nominal values of (500...100) g, to be calibrated using an automatic comparator (which is not equipped with weight support plates). The method can be used for class E_1 weights, where the highest accuracy is required. In this case, the resulting calibration uncertainty for the unknown weights is better than that usually obtained for E_1 masses, being at the level of reference standards. The paper is an update of [2] and brings an improvement of the method described by the author in the paper [3] where the calibration scheme was more laborious.

The article is divided into 7 sections as follows: introduction, equipment and standards used in calibrations, mass determination of reference disc weights used as check standards in the calibration of E_1 weights, statistical tools for evaluation of the measurement process and mass determination of class E_1 weights, analysis of uncertainties, quality assessment of the calibration, conclusions.

4.2. Equipment and Standards used in Calibration

The weighing system includes a proper balance (mass comparator) with weights transporter, a monitoring system of environmental conditions and a MC Link software.

The mass comparator used was an automatic one, with the following specifications:

- Maximum capacity: 1011 g;

- Readability: 0.001 mg;

- Repeatability: 2 to 3 µg in the determinations with Pt-Ir standard and 0.4 to 1 µg in the determinations with stainless steel mass standards.

For accurate determination of the air density an environmental conditions monitoring system was used, consisting in a precise "climate station", model Klimet A30.

Technical parameters for Klimet A30 are:

- Temperature: readability: $0.001°C$; U (k=2) = 0.03 °C;

- Dew point: resolution: $0.01°C$; U (k=2) = 0.05 °C;

- Barometric pressure: resolution: 0.01 GPa; U (k=2) = 0.03 GPa.

The mass standard used for the comparisons was an 1 kg reference standard, Ni81, Fig. 4.1, whose mass value was determined at BIPM.

Fig. 4.1. Reference standard of 1 kg, Ni81.

4.2.1. Some Aspects Regarding the Kilogram "Ni81"

Ni81 had been purchased by the National Institute of Metrology in 1981 and represents the most important standard after NPK. In 2005 and 2013, this mass standard was sent together with NPK to BIPM for calibration and received a new mass value. The first task after receiving these standards back from the BIPM was to perform comparisons between NPK, Ni81 and all reference stainless steel mass standards (a set of two 1 kg mass standards and two sets of disc weights from 50 g to 500 g).

But, from January 2014 to January 2015 an extraordinary calibration using the IPK was carried out at the BIPM. After this calibration campaign, it was concluded that the results obtained for the set of BIPM working standards indicate the existence of an offset from the IPK of 35 µg over 22 years [4]. Therefore, the Consultative Committee for Mass and Related Quantities (CCM) recommended that all mass calibrations of national prototypes and of mass standards issued by the BIPM over the period 2003–2013 need to be amended with this value [4].

Even if the amended mass difference lies within uncertainty values of the reference mass standards (having a magnitude of one half of the expanded uncertainty given in the calibration certificate), it was considered that it was safer to perform additional comparisons at this level of mass dissemination. In this way, as a first stage, three reference kilogram standards were compared directly to the National Prototype Kilogram No. 2, (including Ni81). Then, as a second stage, two sets of disc weights were compared with the stainless steel kilogram, Ni81.

4.2.2. The Weights Involved in Calibration

The weights involved in calibration are:

- Unknown E_1 weights (from 500 g to 100 g, marked with A12...A9) having OIML shape, Fig. 4.2.

- Disc weights (reference weights, marked with NA or NB), Fig. 4.3.

Fig. 4.2. Weights of E1 class.   **Fig. 4.3.** Reference disc weights.

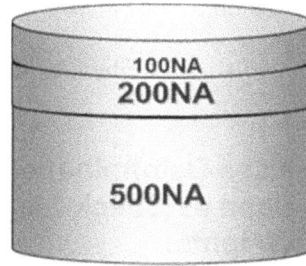

4.2.3. Precision (or Imprecision) of the Balance

Before starting the calibration of the weights, the precision of the balance was monitored using a statistical control technique, according to [5].

A standard deviation of repeated measurements on a single weight was the basis for the test. In this way, it was used a past history of standard deviations on the same balance. If m (in our case, $m=6$) represents the number of standard deviations, $s_1...s_m$, from historical data, a pooled standard deviation was calculated:

$$s_P = \sqrt{\frac{1}{m}\sum s_i^2} = 0.013mg \qquad (4.1)$$

According to [5], the Equation above assumes that the individual standard deviations have v degrees of freedom, in which case the pooled standard deviation has m×v degrees of freedom. For a new series of measurements, the residual standard deviation obtained, s_{new}= 0.0019, was tested against the pooled value. The test statistic, F was calculated:

$$F = \frac{s_{new}^2}{s_P^2} = 1.989 \qquad (4.2)$$

From the Table D.2 of [5], critical value of F is 2.274 for a one-sided test, at the $\alpha = 0.05$ significance level with v=5.

The precision of the balance was considered to be in control because:

$$F \leq \text{critical value from the F-distribution} \qquad (4.3)$$

4.3. Mass Determination of Reference Disc Weights Used as Check Standards in the Calibration of E_1 Weights

4.3.1. Measurement Matrix Design

The first stage of mass unit dissemination is represented in Fig. 4.4, where three reference kilogram standards are compared directly to the National Prototype Kilogram No. 2. Then, in a second stage, two sets of disc weights marked with NA and NB are compared with the stainless steel kilogram, Ni81. The scheme [4] indicates that any reference kilogram can be used for the calibration of disc weights (and further down in the subdivision method for the calibration of E_1 weights), but, in our analysis, was opted for Ni81 as standard.

In this way, were established the possible mass comparisons for the interval 1000 g to 100 g [4] which are presented in Table 4.1. Using the notations from [6], the disc weights were calibrated using the model (1, 1, 1, 1, 1, 1, 2, 2, 0, 2, 2, 0). The weight denoted with 655, which was calibrated in the first stage, acted as check standard.

In the scheme, the standard is represented in yellow whereas the "test" weight(s) is represented in orange.

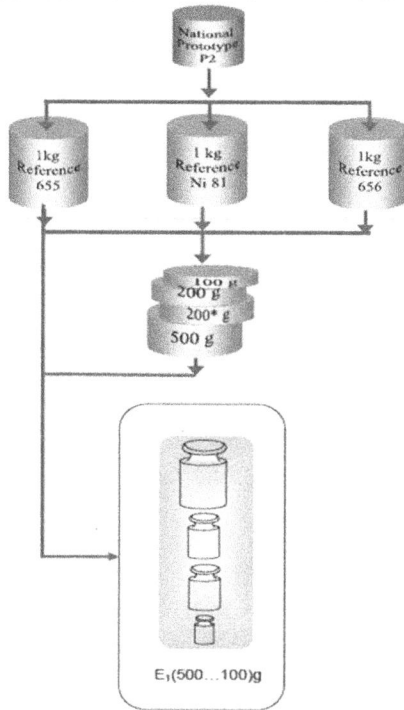

Fig. 4.4. The dissemination of mass unit from NPK to weights of E_1 class.

Table 4.1. Possible mass comparisons between 1 kg to 100 g disc weights.

Det. No.	Mass standard ID							
	Ni81	655	500 NA	500 NB	200 NA	200 NB	100 NA	100 NB
1	−1	1	0	0	0	0	0	0
2	−1	0	1	1	0	0	0	0
3	0	−1	1	1	0	0	0	0
4	0	0	−1	1	0	0	0	0
5	0	0	−1	0	1	1	1	0
6	0	0	0	−1	1	1	0	1
7	0	0	0	0	−1	1	−1	1
8	0	0	0	0	−1	1	1	−1
9	0	0	0	0	−1	1	0	0
10	0	0	0	0	−1	0	1	1
11	0	0	0	0	0	−1	1	1
12	0	0	0	0	0	0	−1	1

4.3.2. Estimated Mass Values for Disc Weights

The theory used for estimating mass values is the least squares approach. The measurement model is:

$$X\beta = Y - e, \tag{4.4}$$

where

β (k,1) is the vector of the k mass values of the standards;

X (n,k) is the design matrix; entries of the design matrix are +1, −1 and 0, according to the role played by each of the parameters (from the vector β) in each comparison;

e (n,1) is the vector of the deviations from the expected values of the mass differences;

Y (n,1) is the vector of n mass differences obtained by weighing (including buoyancy corrections). Each element "y" of the vector is calculated as follows (for real mass values):

$$y = \Delta m + \rho_a (V_1 - V_2), \tag{4.5}$$

where

Δm is the difference of balance readings calculated from a weighing cycle (ABBA);

ρ_a is the air density at the time of weighing;

V_1, V_2 are the volumes of the weights (or the total volume of each group of weights) involved in a measurement.

The expected values of the unknown masses are calculated using formulas [7]:

$$\langle \beta \rangle = (X'^{T} \cdot X')^{-1} X'^{T} \cdot Y' \tag{4.6}$$

The well-known terms from Formula (4.6) are detailed in [7]. The results obtained in this stage, correspond to real mass values. For conventional mass values, the necessary calculations were performed. The conventional mass values and associated uncertainties obtained for this stage of mass determination were included in a calibration certificate. Table 4.2 presents this information only for the disc weights used as check standards in the next step of dissemination:

In Table 4.2, two columns containing calibration uncertainty were introduced: actual uncertainty obtained from the comparisons and that recorded in the calibration certificate (the values covered by the Romanian CMC).

Details on uncertainty components calculation can be found in [4].

Table 4.2. Conventional mass values and uncertainty of the disc weights used as check standards.

Nominal mass	Conventional mass	Uncertainty U obtained from the calibration	Uncertainty U' *according to* "CMC"
500 g NA	500 g + 0.061 mg	0.017 mg	0.035 mg
100 g NA	100 g + 0.003 mg	0.004 mg	0.015 mg

4.4. Statistical Tools for Evaluation of the Measurement Process and Mass Determination of Class E₁ Weights

4.4.1. Method Used to Evaluate the Efficiency of the Weighing Design

The dissemination of the mass scale to E_1 weights, using a single reference standard, requires mass comparisons between weights and groups of weights.

A mass calibration design (or design matrix) describes the general setup of these comparisons.

For a given number of mass comparisons, a criterion for the choice of a design matrix is that, the variances of the estimates be as small as practicable [6].

For this reason, the idea of efficiency was introduced, to enable designs to be analyzed using this criterion, taking into account the variances of the weighing results.

The efficiency is very useful when comparing designs involving the same masses and balances, even if the number of mass comparisons differs. It is desirable that the efficiency of a design be large, as this would indicate that the variances are small [6].

Table 4.3 lists the mass comparisons possible for the 1 kg to 100 g decade, taking into account the following elements: the automatic comparator and the diameter of the disc weights (so that a group of OIML weights can be placed over).

Table 4.3. Possible mass comparisons for E_1 weights
from 1 kg to 100 g decade.

Obs.	Weights ID						
No	Ni81	500NA	500A12	200A11	200A10	100NA	100A9
1	-1	1	1	0	0	0	0
2	-1	1	0	1	1	1	0
3	-1	1	0	1	1	0	1
4	0	1	-1	0	0	0	0
5	0	1	0	-1	-1	-1	0
6	0	0	1	-1	-1	-1	0
7	0	0	0	1	-1	-1	1
8	0	0	0	-1	1	-1	1
9	0	0	0	1	-1	0	0
10	0	0	0	1	0	-1	-1
11	0	0	0	0	1	-1	-1
12	0	0	0	0	0	1	-1

To establish the design matrix "X" of the comparisons, several versions were performed, then calculating the efficiency of the design for each of them.

For example, using the notation of [6], for the design (2, 1, 1, 2, 0, 1, 1, 0, 1, 1, 2, 1) an efficiency of 0.38 was obtained, while for the design (1, 0, 1, 1, 1, 1, 1, 1, 1, 2, 2, 1) the efficiency obtained was 0.61.

Finally, the design (2, 1, 1, 2, 1, 1, 1, 1, 0, 1, 1, 1) was chosen, having 13 equations of condition, since the value for the efficiency was greater, namely 1.04.

The efficiency was calculated in the following manner. Once all weighing are completed, the first step is to form the design matrix, "X", which contains the information on the equations used (the weighing design).

The vector containing the standard deviation of each comparison is represented by "s" and the vector of measured values "y_i" which includes buoyancy corrections, calculated according to (4.12) for conventional mass, is represented by "Y".

$$
X = \begin{bmatrix}
1 & 0 & 0 & 0 & 0 & 0 & 0 \\
-1 & 1 & 1 & 0 & 0 & 0 & 0 \\
-1 & 1 & 1 & 0 & 0 & 0 & 0 \\
-1 & 1 & 0 & 1 & 1 & 1 & 0 \\
-1 & 1 & 0 & 1 & 1 & 0 & 1 \\
0 & 1 & -1 & 0 & 0 & 0 & 0 \\
0 & 1 & -1 & 0 & 0 & 0 & 0 \\
0 & 1 & 0 & -1 & -1 & -1 & 0 \\
0 & 0 & 1 & -1 & -1 & -1 & 0 \\
0 & 0 & 0 & 1 & -1 & -1 & 1 \\
0 & 0 & 0 & -1 & 1 & -1 & 1 \\
0 & 0 & 0 & 1 & 0 & -1 & -1 \\
0 & 0 & 0 & 0 & 1 & -1 & -1 \\
0 & 0 & 0 & 0 & 0 & 1 & -1
\end{bmatrix}
\quad
s = \begin{bmatrix}
0.016 \\
0.0013 \\
0.0013 \\
0.0009 \\
0.0010 \\
0.0017 \\
0.0017 \\
0.0004 \\
0.0013 \\
0.0006 \\
0.0005 \\
0.0005 \\
0.0007 \\
0.0009
\end{bmatrix} mg
\quad
Y = \begin{bmatrix}
-3.1583 \\
3.1896 \\
3.1896 \\
3.0994 \\
3.0758 \\
0.1001 \\
0.1001 \\
0.1796 \\
0.0801 \\
-0.0052 \\
-0.0396 \\
-0.0414 \\
-0.0579 \\
0.0225
\end{bmatrix} mg
\quad
\beta = \begin{bmatrix}
Ni81 \\
500NA \\
500A12 \\
200A11 \\
200A10 \\
100NA \\
100A9
\end{bmatrix}
$$

$$(4.7)$$

The vector β contains all the weights described before. Fig. 4.5 represents a detail of the weights combination: 500NA+200A12+200A11+100A9, part of determination "y_4".

The observations vector Y has a diagonal variance - covariance matrix G:

$$\mathbf{G} = \mathrm{diag}\ (u_r^2,\ s_1^2,\ s_2^2 ... s_{n-1}^2), \qquad (4.8)$$

where u_r^2, is the square of the uncertainty of reference standard, named Ni81, and s_j^2 ($j = 1...n-1$) is the variance of the jth comparison.

If G' is the same as G without the first row and column, then the matrix $G^{-1/2}$ can be calculated.

Fig. 4.5. The combination of the weights from the 4th determination.

By denoting with J a $(n\text{-}1) \times (k\text{-}1)$ a sub-design matrix that would be used if the same mass comparisons are carried out, without the use of a reference mass, the matrix K can be defined:

$$K = G^{,\text{-}1/2} J \qquad (4.9)$$

Calculating K^T, which is transpose of K, one can determine the inverse $(K^T \bullet K)^{-1}$:

$$
\left(K^T \cdot K \right)^{-1} =
\begin{bmatrix}
\mathbf{0.120} & -0.011 & 0.016 & 0.019 & 0.019 & -0.001 \\
-0.011 & \mathbf{0.413} & 0.009 & 0.009 & 0.005 & 0.004 \\
0.016 & 0.009 & \mathbf{0.059} & 0.001 & 0.009 & 0.014 \\
0.019 & 0.009 & 0.001 & \mathbf{0.063} & 0.009 & -0.004 \\
0.019 & 0.005 & 0.009 & 0.009 & \mathbf{0.054} & 0.003 \\
-0.001 & 0.004 & 0.014 & -0.004 & 0.003 & \mathbf{0.071}
\end{bmatrix}
\qquad (4.10)
$$

If v_i are the diagonal elements of $(K^T \cdot K)^{-1}$ corresponding to the ith mass, σ_m is the largest of the σ_i, then the efficiency of the design, represented by the matrix X is defined as [6]:

$$E = \sum v_i^{-1} \cdot h_i^2 \cdot \sigma_m^2 / (n-1), \qquad (4.11)$$

where n is the number of comparisons; h_i is the ratio between the nominal values of the unknown weights and the reference.

Table 4.4 and Table 4.5 present the calculation of the efficiency for different designs containing 13 equations of condition.

Table 4.4. The calculation of efficiency for design:
(2, 1, 1, 2, 1, 1, 1, 1, 0, 1, 1, 1).

$1/v_i$	h	$h_i^2 \cdot 1/v_i$	n-1	σ^2_m	$(h_i^2 \cdot 1/v_i) \cdot \sigma^2_m/(n-1)$	Standard deviation (μg)
8.33	0.5	2.0819			0.501	0.35
2.42	0.5	0.606			0.146	0.64
16.80	0.2	0.672			0.162	0.24
15.82	0.2	0.633	12	2.89	0.152	0.25
18.55	0.1	0.185			0.045	0.23
14.07	0.1	0.141			0.034	0.27
					E = 1.04	

Table 4.5. The calculation of efficiency for design:
(2, 1, 1, 2, 0, 1, 1, 0, 1, 1, 2, 1).

$1/v_i$	h	$h_i^2 \cdot 1/v_i$	n-1	σ^2_m	$(h_i^2 \cdot 1/v_i) \cdot \sigma^2_m/(n-1)$	Standard deviation (μg)
1.946	0.5	0.487			0.117	0.72
1.946	0.5	0.487	12	2.89	0.117	0.72
5.773	0.2	0.231			0.056	0.42
5.222	0.2	0.209			0.050	0.44
8.848	0.1	0.088			0.021	0.34
7.410	0.1	0.074			0.018	0.37
					E = 0.38	

It can be seen that, in the first case (Table 4.4), a higher efficiency was obtained, which indicates that the standard deviations are smaller. Therefore, this weighing design was finally chosen to calculate the mass of the unknown and uncertainty of calibration.

4.4.2. Mass Results Obtained in the Calibration of Weights

If it is denoted by (A) the weighing of the reference weight and (B) the weighing of the test weight, an ABBA weighing cycle represent the sequence in which the two weights are measured to determine the mass difference of a comparison in a design matrix.

The calibration data used are obtained from the weighing cycles ABBA for each yi (which is the weighing comparison according to design matrix "*X*").

The Formula (4.5) for "y", for conventional mass, becomes:

$$y = \Delta m + (\rho_a - \rho_o)(V_1 - V_2) \tag{4.12}$$

with $\rho_o = 1.2 \ kg \cdot m^{-3}$, the reference air density.

To estimate the unknown masses of the weights, the least square method was used [3, 6-7]. The design matrix "X" and the vector of observations "Y" are transformed (to render them of equal variance) in U and W respectively, as follows [6]:

$$U = G^{-1/2} \cdot X \text{ and } W = G^{-1/2} \cdot Y \tag{4.13}$$

$$U = \begin{bmatrix}
0,0625 & 0 & 0 & 0 & 0 & 0 & 0 \\
-0,769 & 0,769 & 0,769 & 0 & 0 & 0 & 0 \\
-0,769 & 0,769 & 0,769 & 0 & 0 & 0 & 0 \\
-1,111 & 1,111 & 0 & 1,111 & 1,111 & 1,111 & 0 \\
-1 & 1 & 0 & 1 & 1 & 0 & 1 \\
0 & 0,588 & -0,588 & 0 & 0 & 0 & 0 \\
0 & 0,588 & -0,588 & 0 & 0 & 0 & 0 \\
0 & 0 & 0,769 & -0,769 & -0,769 & -0,769 & 0 \\
0 & 2,50 & 0 & -2,5 & -2,5 & -2,5 & 0 \\
0 & 0 & 0 & 1,667 & -1,667 & -1,667 & 1,667 \\
0 & 0 & 0 & -2 & 2 & -2 & 2 \\
0 & 0 & 0 & 2 & 0 & -2 & -2 \\
0 & 0 & 0 & 0 & 1,429 & -1,429 & -1,429 \\
0 & 0 & 0 & 0 & 0 & 1,111 & -1,111
\end{bmatrix} \quad W = \begin{bmatrix}
-0,1974 \\
2,4535 \\
2,4535 \\
3,4438 \\
3,0758 \\
0,0589 \\
0,0589 \\
0,0616 \\
0,449 \\
-0,0087 \\
-0,0792 \\
-0,0828 \\
-0,0827 \\
0,025
\end{bmatrix} \ mg$$

$$\tag{4.14}$$

The estimates β_j and their variance-covariance matrix $V_{\beta j}$ are calculated as follows:

$$\langle \beta_j \rangle = \left(U^T \cdot U \right)^{-1} \cdot U^T \cdot W = \begin{vmatrix}
Ni81 \\
500 \ NA \\
500 \ A12 \\
200 \ A11 \\
200 \ A10 \\
100 \ NA \\
100 \ A9
\end{vmatrix} = \begin{vmatrix}
-3.158 \\
0.065 \\
-0.034 \\
-0.053 \\
-0.070 \\
0.005 \\
-0.018
\end{vmatrix} \tag{4.15}$$

$$V_{\beta j} = \left(U^t \cdot U\right)^{-1} = \begin{bmatrix} \mathbf{256} & 128 & 128 & 51 & 51 & 26 & 26 \\ & \mathbf{64} & 64 & 26 & 26 & 13 & 13 \\ & & \mathbf{64} & 26 & 26 & 13 & 13 \\ & & & \mathbf{10} & 10 & 5 & 5 \\ & & & & \mathbf{10} & 5 & 5 \\ & & & & & \mathbf{3} & 3 \\ & & & & & & \mathbf{3} \end{bmatrix} \mu g^2 \qquad (4.16)$$

The diagonals elements V_{jj}, of the $V_{\beta j}$ represent the variance of the weights (which includes the type A variance combined with the variance associated to reference standard).

4.5. Analysis of Uncertainties

4.5.1. Uncertainty of the Weighing Process, u_A

The variance $V_{\beta j}$ can be also expressed as [6]:

$$V_{\beta j} = h\, h^{\mathrm{T}}\, \sigma_r^2 + R \text{ with } R = \begin{pmatrix} 0 & \dfrac{0^T}{\left(K^T \cdot K\right)^{-1}} \end{pmatrix} \qquad (4.17)$$

The diagonals elements of the $(K^{\mathrm{T}} \bullet K)^{-1}$ represents the type A variance of the unknown weight. From here, the type A standard uncertainty can be obtained:

$$u_{A(\beta_j)} = \begin{bmatrix} 0.35 \\ 0.64 \\ 0.24 \\ 0.25 \\ 0.23 \\ 0.27 \end{bmatrix} \mu g \qquad (4.18)$$

4.5.2. Type B Uncertainty

The components of type B uncertainty are:

- Reference standard, u_r;
- Air buoyancy corrections, u_b;
- Sensitivity of the weighing instrument, u_s;
- The display resolution of the mass comparator, u_{rez}.

The detalied calculation of these components can be found in [2].

4.5.3. Combined Standard Uncertainty, u_c

The combined standard uncertainty of the conventional mass of the weight β_j after the calculation of all components is given by:

$$u_{c(\beta j)} = [(u_A^2(\beta j) + u_r^2(\beta j) + u_b^2(\beta j) + u_s^2 + u_{rez}^2]^{1/2} \qquad (4.19)$$

The signification of the terms was explained above.

4.5.4. Expanded Uncertainty

The resulted expanded uncertainty "U" of the conventional mass of each weight, β_j, is:

$$U_{(\beta j)} = 2 \cdot u_{c(\beta j)} = \begin{vmatrix} 500NA \\ 500E1 \\ 200NA \\ 200E1 \\ 100NA \\ 100E1 \end{vmatrix} = 2 \cdot \begin{vmatrix} 0.0085 \\ 0.0085 \\ 0.0035 \\ 0.0035 \\ 0.0020 \\ 0.0020 \end{vmatrix} = \begin{vmatrix} 0.017 \\ 0.017 \\ 0.007 \\ 0.007 \\ 0.004 \\ 0.004 \end{vmatrix} mg \qquad (4.20)$$

The uncertainty components are graphically represented in a "Cause and effect" diagram (Ishikawa) [2, 4], as shown in Fig. 4.6.

4.6. Quality Assessment of the Calibration

As shown, for calibration of the E_1 weights disc weights of 500 g and 100 g were used, having both the role of check standards and weight support plates for the whole determination.

To see if the mass values obtained for disc weights are consistent with previous values, it is necessary to perform a statistical control. The purpose of the check standard is to assure the validity of individual calibrations. A history of values on the check standard is required for this purpose [5]. In the Table 4.6 are presented the results obtained from 10 previous measurements of the two disc weights and also that from the last calibration.

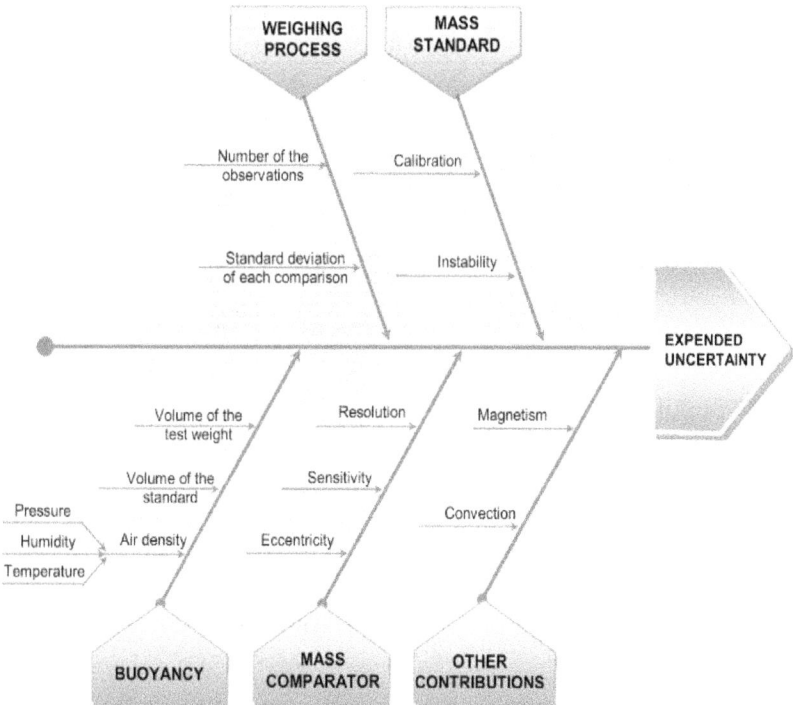

Fig. 4.6. "Cause and effect" diagram.

The accepted value of the mass difference, $\overline{m}_{diff}$, for the check standards, is calculated from the 10 historical data according to Table 4.6. The value of the check standard for the new calibration, m_{diff}, is tested for agreement with the accepted value. For this purpose, the t-statistic, in accordance to [5] is calculated.

$$t = \frac{\left| m_{\text{diff}} - \overline{m}_{\text{diff}} \right|}{s} \qquad (4.21)$$

105

Table 4.6. The measurements results for disc weights.

Year	500 g NA	100 g NA
2007	0.074	0.016
2008	0.067	0.020
2009	0.090	0.013
2010	0.061	0.010
2011	0.057	0.009
2012	0.028	0.001
2013	0.070	0.002
2014	0.065	0.004
2015	0.052	-0.001
2015'	0.061	0.003
$\overline{m}_{diff}$	0.0625	0.0077
s	0.0160	0.0070
2016 (m_{diff})	0.065	0.005

In the Formula (4.21), s is the standard deviation of n historical values of the mass differences, having v-1 degrees of freedom:

$$s = \sqrt{\frac{1}{n-1}\sum_{i=1}^{n}(m_{diffi} - \overline{m}_{diff})^2} \qquad (4.22)$$

Table 4.7 presents the values for t-statistic calculated for the two check standards.

The calibration process is judged to be in control because:

$$t \leq \text{critical value of Student's } t\text{-distribution},$$
with $v = 9$ degrees of freedom

Critical values, can be found in Table D.1 of [5] for a two-sided test, at $\alpha = 0.05$ significance level.

Table 4.7. The calculated t-statistic for the two disc weights.

t calculated		Critical value of Student's t-distribution
500 NA	100 NA	
0.156	0.384	2.262

If for the disc weights there are no sufficient calibration data to perform a statistical control according to [5], the method of normalized error E_n can be chosen, which takes into account the result and its uncertainty from the last calibration.

The results obtained for the disc weights in the subdivision procedure described at chapters 4 and 5, are compared with data from their last calibration certificates [3, 8]. The differences in values are normalized using the Formula [9]:

$$E_n = \frac{\delta_{subdiv} - \delta_{certif}}{\sqrt{U^2_{subdiv} + U^2_{certif}}}, \qquad (4.23)$$

where

δ_{subdiv} represents the mass error of the disc weight obtained by subdivision method;

δ_{certif} is the mass error of the disc weight from the calibration certificate;

U_{subdiv} is the expanded uncertainty of the disc weight obtained in subdivision method;

U_{certif} is the expanded uncertainty from the calibration certificated of the disc weight;

Using this formula, the measurement and the reported uncertainty are acceptable if the value of E_n, is between -1 and +1.

Table 4.8 presents the results obtained for the normalized errors, E_n.

Table 4.8. Comparison of measurement results of disc weights, obtained by subdivision method and results from the calibration certificate.

Nominal mass of disc weight	Subdivision		Calibration certificate*		E_n
g	δ (mg)	U (mg)	δ (mg)	U (mg)	
500 NA	0.065	0.017	0.061	0.017	0.2
100 NA	0.005	0.004	0.003	0.004	0.4

*Even if in the calibration certificate was reported uncertainty according to CMCs, to consider the most defavorable situation, in the Table 4.8 was used for calculation of E_n the value resulted from the calibration of disc weights (see Table 4.2).

Tables 4.7 and 4.8, show the compatibility of the results.

4.7. Conclusions

An evaluation procedure has been presented, used for the calibration of a set of weights by subdivision. This calibration procedure for the determination of conventional mass of the weights is developed in the Mass Laboratory of the National Institute of Metrology, and can lead to an improvement of CMCs (Calibration and Measurement Capabilities), approved and published in the BIPM database.

Before starting the proper adaptive subdivision method, some preliminary measurements were performed: testing the precision of the balance and mass determination of disc weights, which were used as check standards.

The main feature of this adaptive kilogram subdivision method is represented by the fact that the calibration of the weights (whose shape is in accordance with OIML R111) is performed using an automatic mass comparator. Uncertainties obtained using this method for the unknown weights are better than those usually occur for E_1 (when only manual measurements are possible): 0.060 mg for the 500 g weight, 0.03 mg for the 200 g and 0.017 mg for the 100 g, being at the level obtained for reference standards (marked with NA).

Regarding the quality assessment of the calibration, the two methods presented (t-statistic and normalized error) confirm the consistency of the results.

The method described in this paper for calibration of E_1 weights can be used when the highest accuracy is required.

References

[1]. S. Davidson, M. Perkin, M. Buckley, Measurement Good Practice Guide No. 71, *NPL, TW11 0LW*, June 2004, pp. 8-9.

[2]. A. Vâlcu, Improvement of the Calibration Process for Class E_1 Weights Using an Adaptive Subdivision Method, in *Proceedings of the 4th International Conference on Adaptive and Self-Adaptive Systems and Applications (ADAPTIVE 2012),* Nice, France, July 2012, pp. 51-56.

[3]. A. Vâlcu, D. Dinu, Subdivision method applied for OIML weights using an automatic comparator, in *Proceedings of the XIX IMEKO World Congress,* Lisbon, 2009, pp. 281-283.

[4]. A.Vâlcu, The impact of BIPM amendments on Romanian mass dissemination, *Bulletin OIML,* Vol. LVI I I, No. 1, January 2017, pp. 5-10.

[5]. OIML R 111, International Recommendation No. 111, *Weights of Classes E1, E2, F1, F2, M1, M1-2, M2, M2-3 and M3,* 2004.

[6]. E. C. Morris, Decade design for weighings of Non-uniform Variance, *Metrologia,* Vol. 29, 1992, pp. 374-375.

[7]. R. Schwartz, Guide to mass determination with high accuracy, *PTB–MA-40,* 1995, pp. 54-58.

[8]. M. Grum, M. Oblak, I. Bajsić, M. Perman, Subdivision of the unit of mass using weight support plates, in *Proceedings of the XVII IMEKO World Congress,* Croatia, 2003, pp. 407-408.

[9]. International Standard ISO 13528:2015, Statistical methods for use in proficiency testing by interlaboratory comparisons, 2015.

5.

RH Control Developments for Applied Uncertainty Management in Industrial Processes

Alexandra Ionescu and Gheorghe Florea

5.1. Introduction

Process control and optimization represent the current practice for safer and more efficient industrial plants, while risk management represents the starting point for new control algorithms and strategies. There is a stringent need for enhancing plant operations at production management level, because plants often operate near criticality, meaning in conditions far from the ideal ones from the point of view of control and stability. Continuous process industries are usually very complex and difficult to model and keep under control. While plant personnel feel there is a tremendous need for better and more versatile simulation and modeling tools, no product on the market offers the features necessary for dealing with the uncertain nature of complex plants.

Worldwide engineering organizations have developed standards for the engineering of process safety. IEC released two standards IEC 61508, for the suppliers of process safety equipment and IEC 61511, for the end users of process safety equipment. ISA S84.01 "Application of Safety Instrumented Systems for the Process Industry" includes all elements from sensors to final elements, including inputs, outputs, power supply, logic solvers and user interfaces [1].

Alexandra Ionescu
SIS, SA, Bucharest, Romania

Even if we apply these standards we can make plants safer but we do not solve the continuity of production- the main goal in the economic competition.

5.2. Safety and Security

Safety is an important issue nowadays that received an increasing amount of importance lately. This is due to the numerous accidents occurred in industry plants which require the process industry to reconsider the current practices like process design, process control, risk analysis and control, risk assessment. Integrating safety and security with control provides multiple benefits to end-users.

SAFETY ENGINEERING is an applied science strongly related to systems engineering that guarantees that a life-critical system behaves as needed even when components fail. Process safety is the result of a wide range of technical, management and operational disciplines coming together in an organized way, starting with design, continuing with procurement and commissioning and accomplishing with maintenance and then disposal. The tools, techniques, programs etc. required to manage process safety can sometimes be common with those for Occupational Safety (Work Permit system) and in other cases may focus on Process Safety like LOPA (Layers of Protection Analysis) or QRA (Quantified Risk Assessment).

The whole concept of system safety, as a subset of systems engineering, is to influence safety-critical systems designs by conducting several types of hazard analyses to identify risks and to specify design safety features and procedures to strategically mitigate risk to acceptable levels before the system is built.

Additionally, failure mitigation can go beyond design recommendations, particularly in the area of maintenance.

There is an entire realm of safety and reliability engineering known as Reliability Centered Maintenance [2] which is a discipline that is a direct result of analyzing potential failures within a system and determining maintenance actions that can mitigate the risk of failure and involves understanding the failure modes of the serviceable replaceable assemblies in addition to the means to detect or predict an impending failure.

5.2.1. Safety and Security Technologies

Process safety focuses on preventing fires, explosions and accidental chemical releases in process facilities dealing with hazardous materials such as refineries, chemical, oil and gas production and distribution installations. The procedures needed for standardized implementation of safety and security in the control systems should comply with the methodology presented in IEC 61511. The safety life-cycle phases and functional safety assessment stages are presented in the same standard.

For all safety life-cycle phases, safety planning shall take place to define the criteria, techniques, measures and procedures to:

- Ensure that the SIS safety requirements are achieved for all relevant modes of the process; this includes both function and safety integrity requirements;

- Ensure proper installation and commissioning of the safety instrumented system;

- Ensure the safety integrity of the safety instrumented functions after installation; maintain the safety integrity during operation (for example, proof testing, failure analysis);

- Manage the process hazards during maintenance activities on the safety instrumented.

In order to achieve the required level of safety and security, we should take into consideration four important phases: analyze the needed level of SS for the plant, design, implementation and maintenance.

Stand-alone safety systems have been the traditional method of choice, meaning separate design and operation requirements for Basic Process Control Systems (BPCS) and Safety Instrumented Systems (SIS). Separate systems were developed for process control and safety with proprietary operator interfaces, engineering workstations, configuration tools, data and event historians, asset management, and network communications. This approach affects the costs of infrastructure acquisition, plant systems integration, control and instrumentation hardware, wiring, project execution, installation, and commissioning, as well as ongoing expenses such as training, spare parts procurement, and logistics contracts [3]. Until recently, users had little choice other than to use completely different systems for control and safety [4]. "A war of words is raging in the process control industry over the "integration" of safety and control systems. It's a debate that has been ongoing for years, but the recent introduction of new integrated systems by several process controls vendors has lately added fuel to the fire" [5, 8].

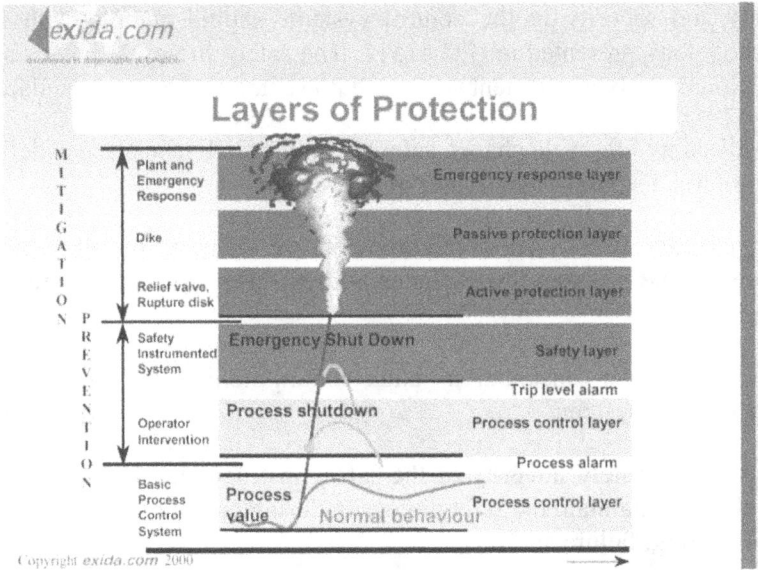

Fig. 5.1. Layers of protection (Exida, 2000).

Today, integrating safety and control has become a cost effective choice for manufacturers that could not justify a separate SIS in the past. As a process manufacturer, you need to perform rigorous Risk and Hazard (RH) analysis based on IEC 61511 or ANSI/ISA-84.00.01 safety standards to decide on the right level of protection required for your plants. You may do that by selecting a SIS that provides close integration with the software tools of your BPCS.

5.2.2. Emergency Shut Down Systems

The Emergency Shutdown System (ESD) is designed to minimize the consequences of emergency situations, related to typically uncontrolled flooding, escape of hydrocarbons, or outbreak of fire in potentially explosion areas but also other kind of areas which may be hazardous.

An emergency shutdown system for a process control system includes at least detectors or sensors, emergency shutdown valve and an associated valve actuator. An emergency shutdown controller provides output signals to the ESD valve in the event the process control system can not act to prevent the shutdown.

Typical Actions of an Emergency Shutdown System:
- Shutdown of systems or parts and equipment;
- Isolate explosion risk inventories;
- Isolate electrical equipment;
- Prevent escalation of events;
- Stop explosion risk flow.

5.2.3. Fire and Gas (F&G) Detection and Alarm Systems

In most industries, one of the key parts of any safety plan for reducing risks for personnel and plant is the use of early warning devices such as gas detectors. These can help provide more time in which to take remedial or protective action.

Instead ESD, a preventive layer of protection, an F&G is a mitigating layer of protection because the purpose is to reduce the consequence severity of such event. When a combustible gas, a toxic gas, flame or heat is detected then the F&G will respond by annunciating sound and/od visual alarms and initiate fire suppression systems, water deluge or even shut down process. In the event of gas leak F&G System must act to prevent it to become a fire or explosion by isolating the leak and ignition sources.

They can also be used as part of a total, integrated monitoring and safety system for an industrial plant.

5.2.4. Burner Management Systems

A Burner Management System is a safety solution for burner equipment such as boilers or furnaces that enables the safe start-up, operation and shut down of the multiple burns. It reduce maintenance, improves up-time and provide safe environment for the equipment and personnel.

Main components of the system are emergency safety shut-off valves, flame monitoring and gas detectors, and BMS controller. Each burner is equipped with its own flame monitoring devices, independent detectors are required to supervise the pilot and the main flame. In most installations two different detection techniques are used, a flame (ionization) rod to monitor the pilot flame and an UV (Ultraviolet) or IR (Infrared) to monitor the main flame.

BMS systems are based on a Programmable Logic Controller (PLC) as the primary safeguard and logic solver, SIL certified.

5.3. RH Control the New Level of Decision

5.3.1. Generalities

In order to achieve the required level of safety and security, we should take into consideration four important phases: analyze the needed level of SS for the plant, design, implementation and maintenance.

You may do that by selecting a SIS that provides close integration with the software tools of your BPCS while still providing the required degree of separation. Fig. 5.2 illustrates three options.

In a traditional sense, process safety refers to additional components that protect personnel and plant from injury, death and economic loss. However, many end users now recognize that the deployment of intelligent integrated safety solutions can directly improve process and personnel safety.

The entire issue of safety has direct influence upon the activity of the plant and therefore it must be integrated into the control system.

According to process safety standards, the process risk has to be reduced to a tolerable level as set by the process owner [6]. The solution is to use multiple layers of protection, including the BPCS, alarms, Operator Intervention (OI), mechanical relief system and a SIS.

The BPCS is the lowest layer of protection and is responsible for the operation of the plant in normal conditions. If BPCS fails or is incapable of maintaining control, then, the second layer, OI, attempts to solve the problem. If the operator also cannot maintain control within the requested limits, then the SIS Layer must attempt to bring the plant in a safe condition. If SIS also fails in restoring normal operation, then the hazard is imminent.

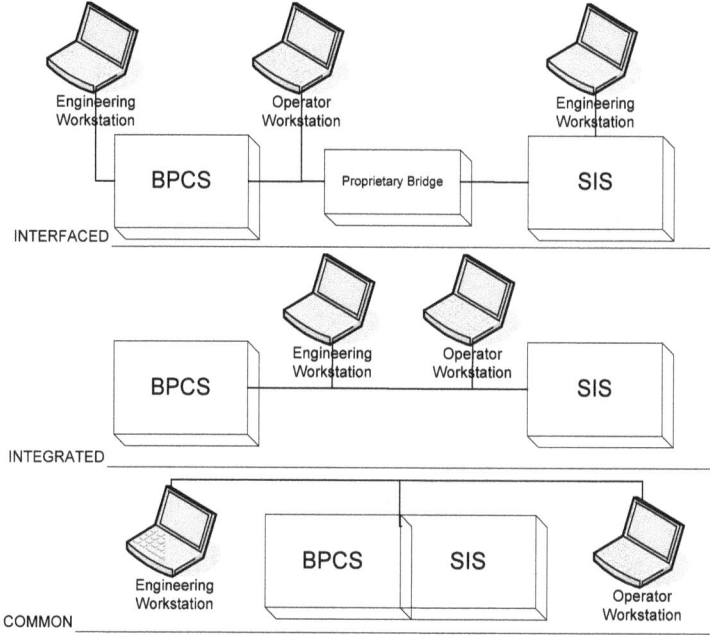

Fig. 5.2. SIS and BPCS Integration Levels.

Even there is still actual the debate between integrated or separate systems [7] more companies adapt their standards to the technology evolution and international standards.

Risk is defined as the combination of the probability and the severity of a hazardous event, meaning how often it can appear and how severe are the consequences when it does. The best way to reduce risk in plant is to design safer processes. Unfortunately, it is impossible to eliminate all risks, so a manufacturer must agree on a level of risk that is considered tolerable. After identifying the hazards, a Risk and Hazard analysis must be performed to evaluate each risk situation.

Risk assessment (Fig. 5.3) is the first process in the risk management methodology to determine the extent of the potential threat and the risk associated with a system [9].

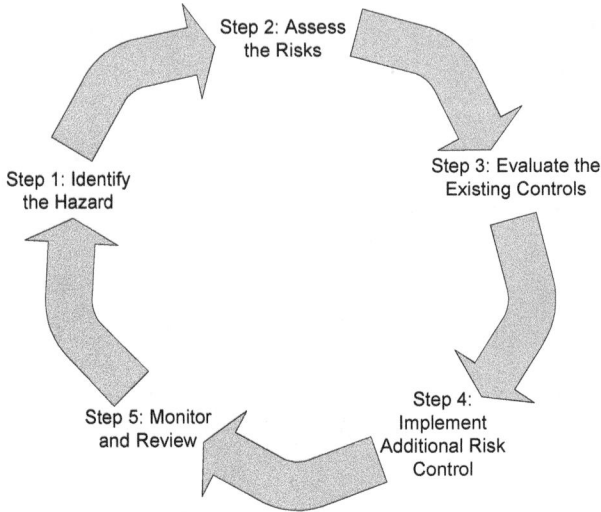

Fig. 5.3. Risk assessment procedure.

5.3.2. RH Control – the New Challenge

BPCS, along with process alarms and facilities for manual intervention, provide the first level of protection and reduce the risk in a manufacturing facility. Additional protection measures are needed when a BPCS does not reduce the risk to a tolerable level. They include SIS along with hardware interlocks, relief valves, and containment dikes.

All this can help only to safely shut down the plant. We [10, 20] have proposed, designed and implemented a new level of decision: RH Control (Fig. 5.4) to keep running the plant.

Better automation is a key aspect for improving industrial competitiveness. Intelligent automation, at management levels in particular, can play a major role regarding this aspect. The purpose of RH Control is to help with this improvement by building a new architecture and a distributed, generic decision support software system for near critical situation management in continuous process industries. In particular, assistance in terms of diagnosis and elaborating solutions is provided (directly to the plant's control system and/or to the staff) when certain situations are detected, i.e. situations suitable to be corrected, prevented or enhanced.

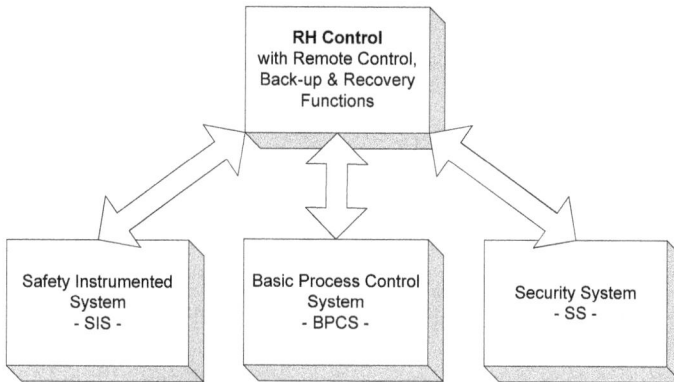

Fig. 5.4. System architecture.

The focus is on new algorithms and strategies for the integration of different software components as well as on the system architecture itself. These software components include core, user interface and problem solving modules.

RH Control follows the conceptual structure of most distributed control systems that is a hierarchical and multilayered structure, similar to a pyramid. The complexity of the control mechanism increases for the higher layers. All the basic functionalities of the system are grouped into problem solving components that work in a cooperative way to find a solution to the plant problems or to optimize according to the plant objectives.

These applications include the following functionalities at different control layers:

- Strategies: Management of global objectives of the plant and their interrelation (management of maintenance operations, incident prevention, RH control, assessment of production costs in real time, loop tuning optimization, set-point deviation detection and alarm management).

- Tactics: Assistance through the problem lifespan, including process failure prevention, risk detection and diagnosis, plant-wide analysis, corrective actions, actions or recommendations for reestablishing effective control.

- Operations: Tasks such as filtering and validation of plant data, variable estimation, alarms analysis and optimization, intelligent alerting based on intuitive technologies and trend forecasting.

5.3.3. Control and Strategy

The requirement of improved reliability, efficiency and maintainability during process operation brings more challenges to the process control design. Traditionally, the prime control objective has been to maintain desired process performance while ensuring robustness against process disturbances. With such design system, in the event of critical faults, the complete control performance can either deteriorate significantly, or may even collapse. It is well known that customer requirements must be translated into design specifications for products and services. Specifications include design optimums or targets, as well as limits which define the minimum or maximum of given characteristics that a company wishes to deliver to its customers.

Control Strategy is a planned set of controls, checks and sequences, derived from current product and process, having as target to insure process performance and product quality. The controls can include facility and equipment operating conditions, product specifications, in-process controls, process parameters, attributes and circumstances related to risk and hazard and the associated methods, frequency of control monitoring and control compliance. Therefore, RH control is the ability to constrain process variation and prevent nonconformance over time, maintaining the stability and functionality even in hazard and risky situations.

Based on the introduction of a new decision level- Risk and Hazard Control and a new state of the process- safety state, the layers of protection and also the impact over the process are changed accordingly (Fig. 5.5).

RH Control according to our new approach aims to solve this problem by building a new architecture that addresses decision support for near critical situation management in continuous process industries. Assistance, in terms of diagnosis and solutions, is provided to the plant and/or to the staff when situations suitable to be corrected, prevented or enhanced are detected. The goal is to conduct the process to the safety state by RH control strategies [17]. Fig. 5.6 illustrates the process state diagram.

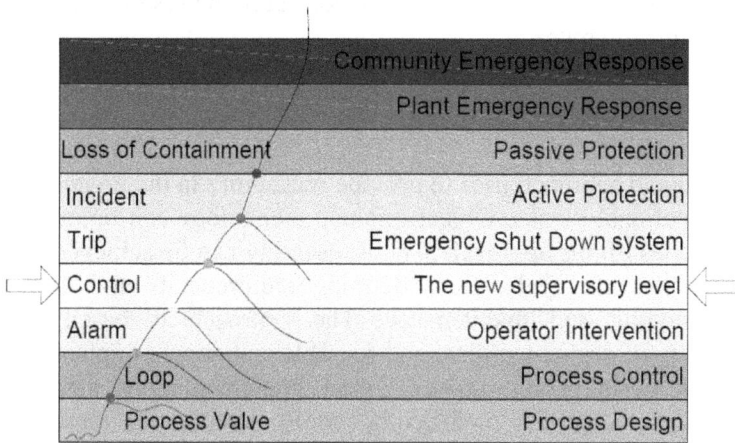

Fig. 5.5. Layers of protection and impact on process.

RH State is a provisional process state introduced for a very short time if the control strategy is not able to maintain it and the shutdown is inevitable. If in this interval RH control is successfully performing, after some time the process can be driven to the nominal state. Time components can be easily changed without affecting the behavior of others.

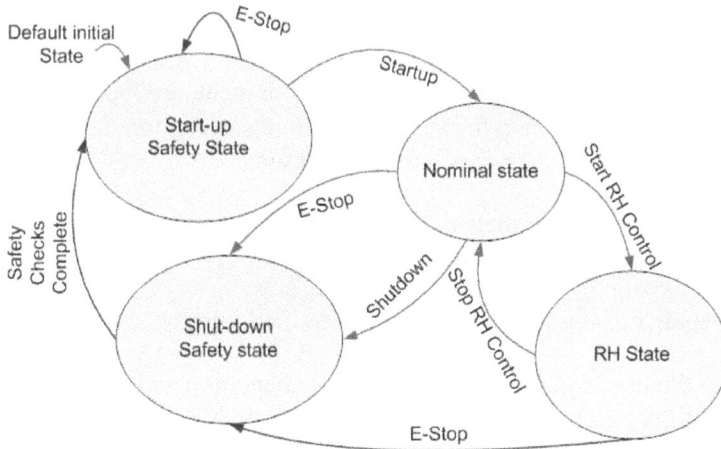

Fig. 5.6. Process states.

The main challenges when we start to configure the system are software architecture and reusability.

5.3.4. Reusability

The technical approach tries to provide reusability in the broadest sense using functional blocks. Object oriented technology can be one of the cornerstones of this approach [11]. Reusability can be achieved for any stage in the life cycle: from defining requirements and design to commissioning and maintenance. The approach is based on the availability of design template and reusable component implementation with few design compromises. These implementations are flexible enough to be adapted or modified to comply with the new requirements with little effort. The concepts of function block based development and integration middleware provide the basis for reusability. RH Control will incorporate components for process control, risk analysis, optimization, etc.

The customized components will be integrated in a global architecture using real-time integration. This software, based on function block philosophy, will incorporate extensions to make possible for its use in real-time applications. This facilitates the easy reuse of components and even of the global application architecture.

5.3.5. Software Architecture

The software architecture is Service Oriented (SOA) [12]. In most applications, infrastructure and the environment are very important security-related issues in the system and it gets even more important if a SOA based on Web Services has been chosen.

For this purpose, asymmetric cryptography will be used, implying a pair of two keys: public key and private key.

The benefits of this approach can be classified into two categories:

- From the user's point of view: the implementation addresses problems related to the global management of the plant while taking into account the interrelation of the strategic objectives, such as production, quality, maintenance, safety, efficiency and availability, as well as problems closer to the process control layer.

- From the systems integrator's point of view: the development of an open software architecture, based on the OPC standard and function blocks, will allow the construction of distributed intelligent control systems on top of the existing ones, with back-up functions.

5.4. Emerging Technologies

The emergent technologies [13] used for the design and achievement of control, safety and security integrated systems can concur substantially to supporting the designers in materializing the proposed architectures. Advanced computation capacities contribute to realizing increased economic efficiency means, which determine not only the theorists to be preoccupied, but mostly the practitioners that seldom obtain important results using adequate methodologies and techniques. Among these, the simulation, concurrent engineering and asset management technologies distinguish themselves lately through important development, due mainly to the evolution of IT&C resources.

5.4.1. Simulation Technologies

Simulation in the field of process control or risk analysis is not something new, the achievements being mainly theoretical. The evolution of information technology, the hardware and communication performances and the software capacities are multiple, but the most important used in this way are the following:

5.4.2. Developed Computer Networks

The performances in the field of computer networks allow the cumulating of the computational capacities of several computers at high speed and the parallelization of data processing in a way that makes possible the simulation of complex systems and the high accuracy connection of the control system with the simulated model.

5.4.3. Intelligent I/O Interfaces

The classical microprocessor interfaces were used for validating, selecting and memorizing the inputs and outputs. They are now replaced by intelligent interfaces capable to prefigure the evolution of outputs at frequencies superior to that necessary for the simulation of the controlled process, insuring thus the predictability of the process evolution.

5.4.4. High Speed Simulators

These are based on the Hardware-in-the-loop (HIL) concept, used especially for testing embedded systems, through which real components and/or controllers are introduced in the simulating model. A larger approach is that of the Hybrid Process Simulation (HPS), which is a testing procedure that combines at least a simulated component and a real one, frequently used for developing virtual manufacturing systems.

5.4.5. The Systems Modeling and Simulating with Discrete or Hybrid Events

Most of the studies referring to dynamic systems modeling, analysis and control are focused on continuous dynamic processes, which can be described with the help of differential equations. Another view is that of discrete events systems, which is applied especially for controlling high level hierarchical systems, planning and giving an order to complex process activities or communication systems. In practice, there are numerous systems made up of continuous subsystems but also of discrete subsystems, and their interaction is vital for the global behavior of the system. This type of systems is named hybrid dynamic systems and for their simulation numerous ways of modeling are used. We can mention the status discrete diagrams, finite automata, Petri networks and Markov chains.

5.4.6. On Line Testing and Diagnosis

In the case of critical processes, it is recommended that the testing be executed in simulating mode, because this way the functionality of the system is not affected, avoiding the high risk situations. The on-line coupling of the simulator with the controller, without physical connections, allows the concomitant realization of the control and test strategies and allows the possibility to detect the flaws during the process operation. An adequate model of the process that works concomitantly allows the comparison of the virtual commands with the real one, contributing decisively to flaw detection (see Fig. 5.7).

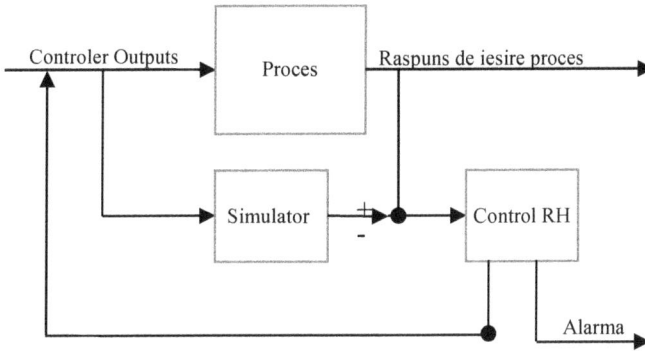

Fig. 5.7. On-line diagnose.

5.4.7. Asset Management

The new approach proposed for the asset management, through which there can be connected not only the process physical resources (machinery, equipment, pipes, cables, buildings, foundations, stockades) but also those of the control systems (linked to instrumentation, safety, security, communication, information technology) is dedicated to increasing security in functioning and insuring the systems functioning attributes in uncertain conditions. The evolution of the integration mode of the technological installations with the automation equipment will continue by incorporating the human operator, maybe the best asset in insuring the functioning and the security of the installation. New asset management approach will contribute to risk and hazard reduction based on preventive and proactive maintenance. Big data exploitation will permit to develop deep diagnosis procedures and will contribute to earlier intervention.

5.5. System Development Methodologies and Techniques

5.5.1. System Engineering

Design is the process by which human intellect, experience, creativity, expertise, and passion are translated into useful artifacts. Engineering design is a subset of the design process in which performance and quality objectives and the underlying science are particularly important. Engineering design is a structured and methodological activity that includes problem definition, learning processes, representation, and decision making.

Engineering designers must develop solutions to satisfy particular specifications while complying with all constraints, standards and regulations. The traditional design approach has been one of deterministic problem solving, typically involving efforts to meet functional requirements subject to various technical and economic constraints.

A number of approaches have been proposed to organize, guide, and facilitate the design process seeking a logical and rigorous methodology to aid in developing a satisfactory design, or one that is acceptable to the user of the product. Examples include Taguchi's theory of robust design, Deming's principles of quality control, Quality Function Deployment, design for manufacture, and concurrent engineering. It is important that this approaches be assessed individually and collectively to determine both their strengths and limitations for particular applications.

The role of decision making in an engineering design context can be defined in several ways. As shown in Fig. 5.8, the decision process is influenced by sets of conditions or contexts.

The business context represents the target of the engineering company and is under his control but the environmental context, such us politics or economy is not controlled. The input context done by variation and changes in requirements and constraints, is established by the customers. The output context is based on the capacity to implement decisions, risks, and qualifiers. Decision Tools and Processes, the level of technology but also the companies culture, data management system and team expertise are arguments to respond to customer expectation and evolution of economy, standards and regulations and implicitly to completion. It is extremely important that engineering company are able to work with other design companies or customers worldwide and this goal is achieved only based on computer tools such as; knowledge-based engineering, workflow, and collaboration. This tools provide computational procedures of engineering design rules, allowing engineers to execute modeling, analysis, and optimization. Collaboration tools, including Internet based conferencing and graphics sharing, help day-to-day working relationship between engineers at distributed locations [14].

System engineering theory helps designers create a hierarchy of functions and physical objects.

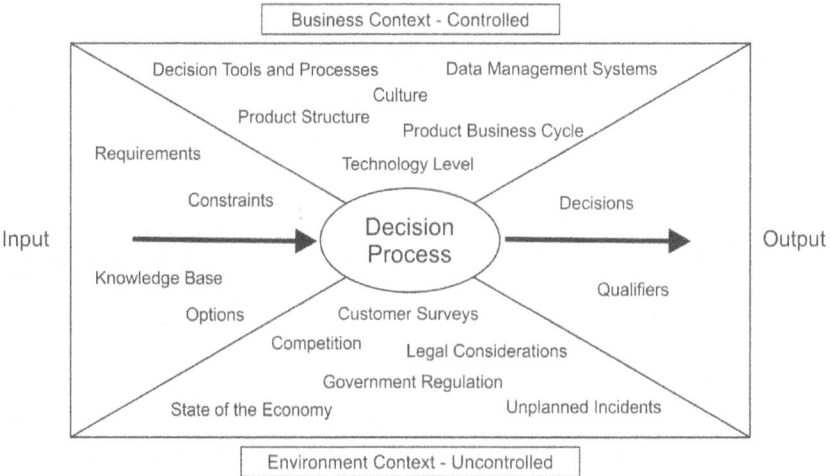

Fig. 5.8. Decision process in the context of business and environment (National Research Council, 2001).

During system design, the requirements of upper levels in the hierarchy are decomposed and flowed down to the lower levels to create separate manageable pieces that can be worked on independently. Major challenges include remembering all the requirements, keeping them consistent, and understanding the many interactions between branches of the hierarchy. These interactions can cause problems during integration at the end of product development.

There are some very important stages; System Architecture and Software Architecture that must help to avoid such problems. In the meantime a methodology that comprise iterations using modeling and simulation for parts or ensemble in the Factory Acceptance Tests(FAT) with the direct involvement of User can assure the compliance between user needs, level of technology and performances. System engineering stages (Fig. 5.9) help the developer to organize his work in order to achieve the stakeholders requirements.

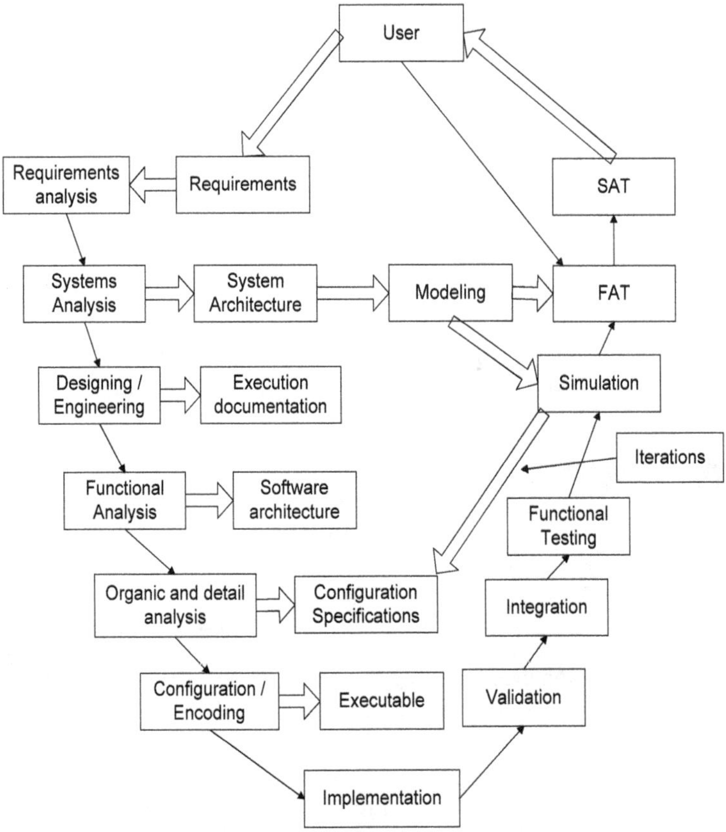

Fig. 5.9. System Engineering stages.

5.5.2. Extended V Model for Uncertainty Control Development

There are a lot of approaches developed for systems engineering and control architecture engineering. The most suitable seems to be V Model that is now widely used including a lot of industries [15]. Our goal is to demonstrate that this model extended with two wings can be the best model to be applied for the development of control systems to be able to work under uncertainty. To represent his type of systems is most suitable because in the left wing is comprised the use of a families of architectures or an adequate framework and in the right wing is provided the capability of the system to evolve, by improvements or up-grades in the entire life-cycle till the retirement.

The most important step in system engineering is the architecture development. It is identified the model of the architecture or the part that is applicable. Artefacts of planning, programming, procedures, frameworks are identified and established to be followed and developed. In this stage there are to be:

- Defined the project area where will be applied;

- The structure envisioned or the architectures classes to be used, including the integration of existing structures, modules, components opportunities;

- The possible improvement of the framework with the implementation capabilities;

- The planning and project development in order to respect the time schedule.

The engineering process comprise key activities such as:

- Identification of the architecture/architectures suitable to be applied;

- Identification of the architecture parts to be used;

- Check of the project consistency related to the propose architecture;

- Identification of the necessary changes and solutions to be used.

Extended V Model (Fig. 5.10) for product development can be applied to control development for facing to uncertainty. The main reasons for this approach are based on our experience and can be summarized:

Left wing:

- Always there is a class of used and verified system architecture and process life cycle start with choosing to be the framework one of them;

- Concept exploration must accommodate the user requirements with control system feasibility.

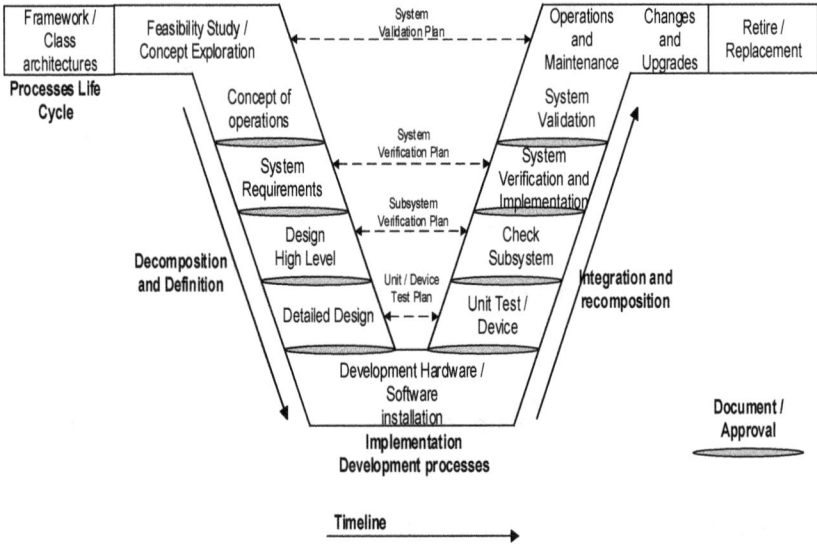

Fig. 5.10. Extended V Model applied to systems development.

Right wing:

- The system must be open to changes and up-grades to react to environment changes including uncertainty;

- The system must accomplish his functions and goals till retirement, based on his attributes.

5.5.3. Architecting Systems

Within the architecting process, several types of architectures are involved [21]. These are:

> The functional architecture (a partially ordered list of activities or functions that are needed to accomplish the system's requirements). The objective of this approach is to:

- Define the elements, activities, tasks, data and information exchange;

- Embed modules with standard functions and tasks;

- Description of generic activities non related to organizational model.

> The topological architecture (at minimum a node-arc representation of physical resources and their interconnections). The objective of this approach is to:

- Define the operational tasks correlated with each component resources;

- Define the interfaces in order to achieve interconnectivity;

- Colligate the components with functions, characteristics, performances and interconnections.

> The technical architecture (an elaboration of the physical architecture that comprises a minimal set of rules governing the arrangement, interconnections, and interdependence of the elements, such that the system will achieve the requirements). The objective of this approach is to:

- Colligate the operational and functional tasks correlated with each component construction based on high technology;

- Define standards and rules applicable but also procedures for development, manufacture, implementation, exploitation and maintenance;

- Ensure information and data exchange with content, speed and performances impose by the process.

> The dynamic operational architecture (a description of how the elements operate and interact over time while achieving the goals).

> The holonic architecture (a distributed, heterarchical structure based on holons with autonomy, cooperation and holarchie).

5.6. Concurrent Engineering

Changing from the classical approach of the software engineering, as part of the complex automation systems design methodology, to the concurrent one comes from the need to insure simultaneously the predictability of costs and development time, together with meeting the quality criteria. The main objectives of applying concurrent engineering [16] during the architectural design phase are:

- Shortening the time for the system accomplishment;
- Increasing profitability;
- Raising competitiveness;
- More control on design and production costs;
- Improved and guaranteed quality;
- Visibility and promotion of the new design philosophy.

The international system design community is still using old methods due to the fact that the organizational complexity is still hard to understand, a large number of specialists being needed in order to simultaneously work on many components and their interaction. Managing this process calls for a clear semantic formalism, the knowledge of the mathematical model being essential to the analysis and the profound understanding of this development process. Another model is also needed in order to execute, assist or coordinate the development process. Compared to other research fields, software engineering passes through a period of quick and profound changes regarding the methods, media and procedures. The modern process of software development has three fundamental characteristics: short but numerous life cycles, high level of simultaneity, behavioral unpredictability (lack of determinism). These characteristics find themselves in contradiction with the work technology that enforces that the process be ordered, predictable and repeatable. A typical model of a classical work flow stands on the following premises:

- A process is a series of steps that are to be executed as they follow, thus the execution of the last step meaning ending the process and accomplishing the objective;

- The processes are repeatable step sequences;

- The product that responds to all the process objectives is taken into account, the next versions being only improvements of this product.

In software engineering, the process is not the same thing, because:

- The final objective is not always reached, the actions lacking determinism;

- The concurrent processes are executed in significantly different ways;

- The product is not considered finalized when reaching the objectives of the process (which is not always guaranteed), each version being saved as an important aspect for improving performance;

- The versions of the product are continuously and in parallel modified.

Software design in the concurrent engineering is by excellence a cooperating process in which a group of specialists with collective tasks work in the same time, but coordinated with the same objective of a unique product development. Simultaneity can't be avoided in the cooperating processes, because the simultaneous access to resources and data can't be restricted, but it can lead to the increased number of no convergent situations that imply the risk of obtaining an inconsistent result. On the other hand, the lack of simultaneity leads to a big waste of time and unpermitted delays for a competition market. The dilemma that the concurrent engineering needs to solve is to find the optimum compromise between high speed and implicitly high concomitance, and respectively low risk and implicitly low concomitance. The objective of the concurrent engineering is to find means of increasing simultaneity limiting in the meantime the risks, and this thing can be obtained by implementing methods to insure the convergence of the solutions.

This approach, needing an increased design effort, has immediate results because the savings in achieving the system/process and especially its efficiency due to reaching the planned performances are remarkable. The paradigm of the concurrent design of system architecture and of process control is based on the principle that the controlled process and the control system are designed in parallel before the construction of the installation. Within the BIOSIS project, this method was approached for the first time in Romania, based mainly on the achievement of the risk and hazard HAZOP study during the design phase, study that influences in this way the design from the very first stage of the technological installation. Also, the control and shutdown system (ESD- *Emergency Shut Down*) will insure the entire safety of the process.

The analysis of the interactions and interdependencies between technology and control is done in advance, thus avoiding conflicts and jams of the two design activities. In this way, underperforming and useless activities, processing and functions can be eliminated, upstream and downstream from the installation, from the preliminary stages of the design. Aspects regarding the process control, its stability, the safety and security in exploitation can be approached and made more efficient in same time. Partial approaches can be found in the specific literature, for example regarding the controllability analysis during the first stages of design]. Although we cannot yet find results of the research of integrated approach of concurrent engineering for achieving technological installations in the main time with the control system, some results are available and we can find products that can be used for the design of a frame structure for the demonstration and development of the concept, including modeling, simulation, on-line identification, risk and hazard analysis, asset evaluation. In this context, our approach [17] answers the need to have at our disposal even from the first stages of design the way to optimize the operations downstream, including the capacity to control the functioning of the process during risk and hazard situations. The concurrent engineering means important financial efforts, but that can be compensated by the reduction of realization costs. His paradigm is based on the principle that not only the process architecture but also the associated control strategy are designed in parallel before the realization of the objective. The interdependencies analysis is done before the design of the two entities (process – control) based on conflict solving criteria. Even if it is based on HAZOP studies that theoretically solves the problems of the systems' functioning in uncertain conditions by implementing the SIS systems, the concurrent engineering must be applied to the process -control, safety and security integrated system ensemble as a whole. The evolution of software engineering methodologies, from a linear approach to that in spiral and afterwards to an agile one show that the iterative development in short cycles and a strong competition are key factors in top software engineering. Using the concurrent engineering is recommended not only for finalizing the global system architecture, but also (or mainly) for developing the software architecture.

An activity that requires a high degree of effort from a design company, but not without a rewarding return on investment is CE.

This design paradigm is based upon the principle that the process and the associated control strategy are designed in parallel before the process is built. Trade-off analysis is performed in advance, in order to prevent conflicting criteria of the two designs. Dynamic process simulators are combined with traditional static simulators to assess transient behavior and controllability of the process. Fig. 5.11 presents a principle scheme regarding the modality of applying the concurrent engineering concepts.

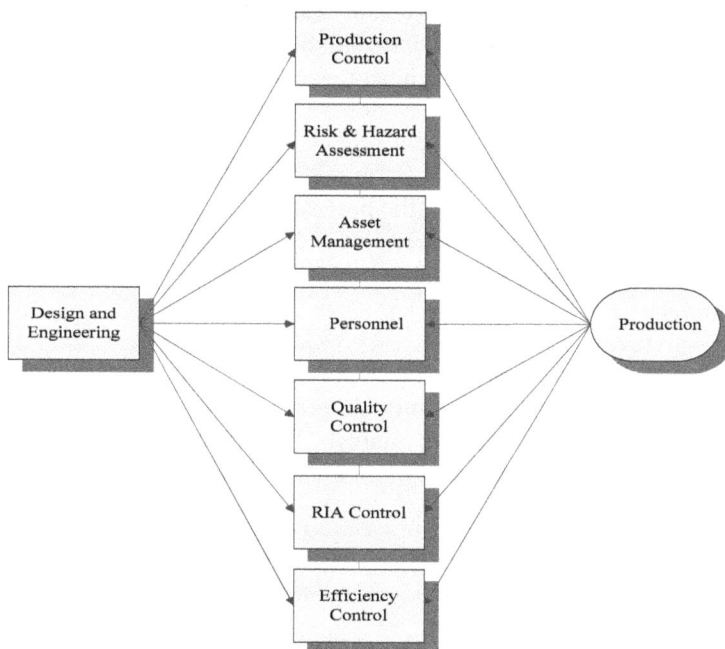

Fig. 5.11. Concurrent engineering apply to process control.

The CE approach is based on the following key elements:

- The system engineering process;

- A multidisciplinary team (process, control, safety and security, management, accounting, inventory);

- A collaborative platform, control environment and data & information distribution;

- Supporting tools and facilities.

The approach can evolve into an Integrated System Development based on cross functional System / Process Teams for all systems and services, and a System Engineering and Team to cover the system issues, performances, balance requirements.

By applying CE to plant design and installation, non-value added activities both in the upstream and downstream activities of the plant can be eliminated at the early stage of the design process, plant, operations and control [17]. Plant wide controllability analysis in the conceptual design stage is an issue that has been raised by process industry [19].

The role of CE is obvious since it reflects that at the conceptual design stage, opportunities exist to optimize the downstream operations including the capabilities to run the process instead of risks and hazards. *This is against the conventional approach of the control as an add-on to process design after the flow sheet structure has already been determined.*

There are a number of tools available for the design of process using CE including: simulation, process modeling, on – line identification, asset assessment, and risk and hazard analysis. Including all this we can have a conceptual framework (Fig. 5.12) for the implementation of CE in process control.

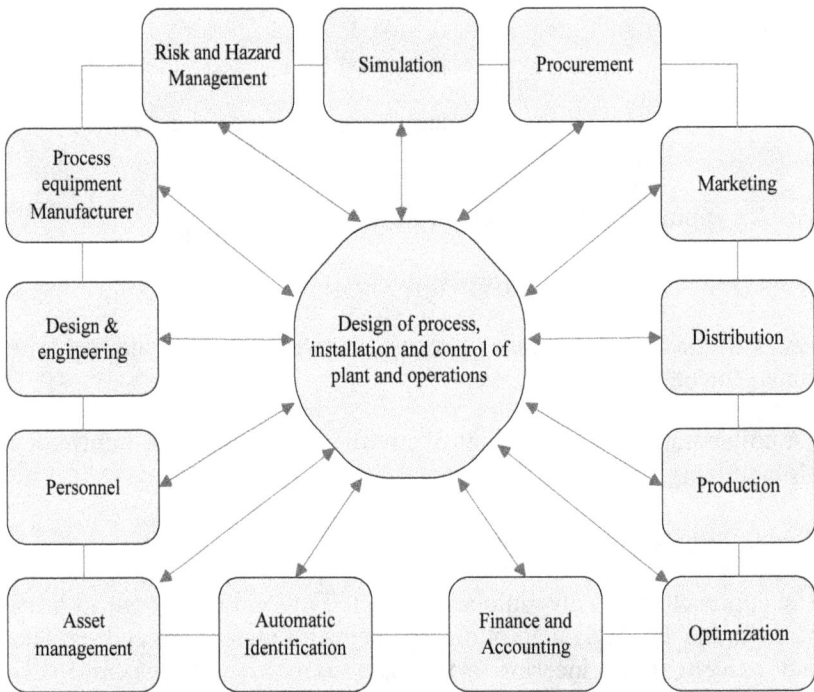

Fig. 5.12. Conceptual framework.

5.7. Applying Uncertainty Management Principles for System Architecture Design

5.7.1. A Framework for Control System Architecture Design

Research in software technology has the potential to revolutionize control system design and implementation. Component-based architectures encourage flexible "plug-and-play" extensibility and evolution of systems. Distributed object computing allows interoperation and dynamic reconfiguration is feasible based on technology and standardization advances enabling the evolution of systems while they are still running. Technologies are being developed to allow networked, embedded devices to connect to each other and self-organize.

To develop a framework for control system design we propose the architecture presented in Fig. 5.13.

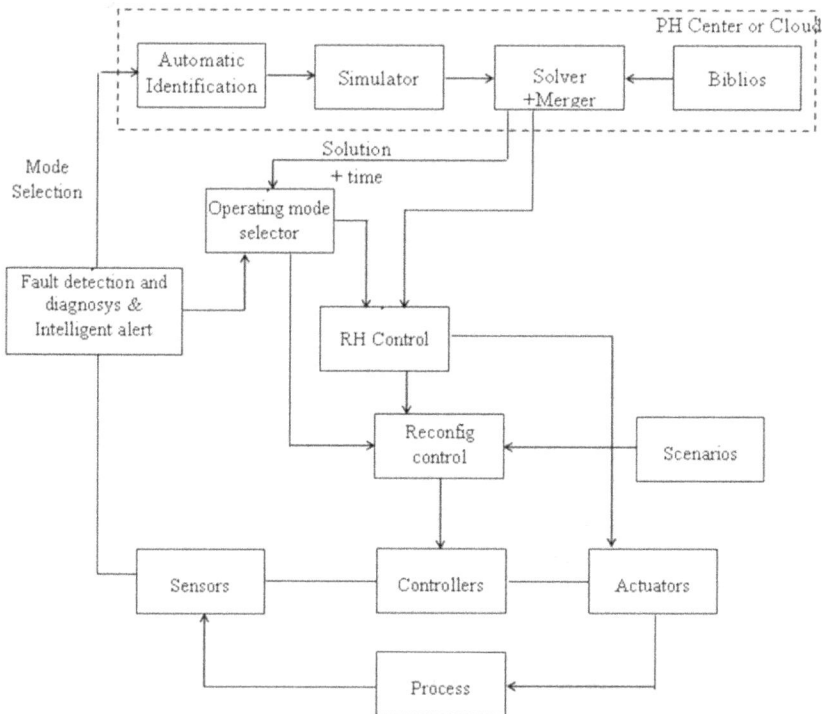

Fig. 5.13. Reconfigurable control development framework.

137

The framework is able not only to host the entire process control system (hardware, software, application, operator and engineering interfaces) but also the model, even simplified, of the process, the simulation features, a library of algorithms and strategies, case studies. The focus in our work is on developing a structure of fault detection and intelligent alert that in conjunction with RH control can conduct to the recovery of functionality, even with spoiled performances. Mode selection part of this structure functions as follows: first, the fault recovery measures for individual loop failures are derived from a fault impact analysis, next, the fault recovery principle initiates a change in the operating strategy of the plant by incorporating changes in the operating factors associated with failures in the model based control calculations. These strategies can be implemented with direct commands from RH Control or/and associated with reconfiguration scenarios.

5.7.2. A Holonic Architecture for Plant Wide Control

Depending upon the failure type, a typical control problem reformulation would involve one or more the following tasks:

1. Modifying set-points;
2. Redefining constraints/limits;
3. Changing the internal model to reflect the fault condition;
4. Changing the control strategy according with safety state.

The results are taken in the design of the process control architecture. Integration of RH control able to maintain the process in the safety state with control hierarchy layers is shown in Fig. 5.14, where A, B, C and D suggest a holonic organization at several levels.

To be able to perform such tasks, the system architecture, structure and data flows must be able to support different methods of reconfiguration. Consequently, reconfigurability design must focus on:

1) Components (Sensors, Actuators, Controllers, Configuration);

2) Control (Algorithms, Structure, Data flows, RH control strategies, Integrated control);

3) Process (Equipment, Flows, Process, States).

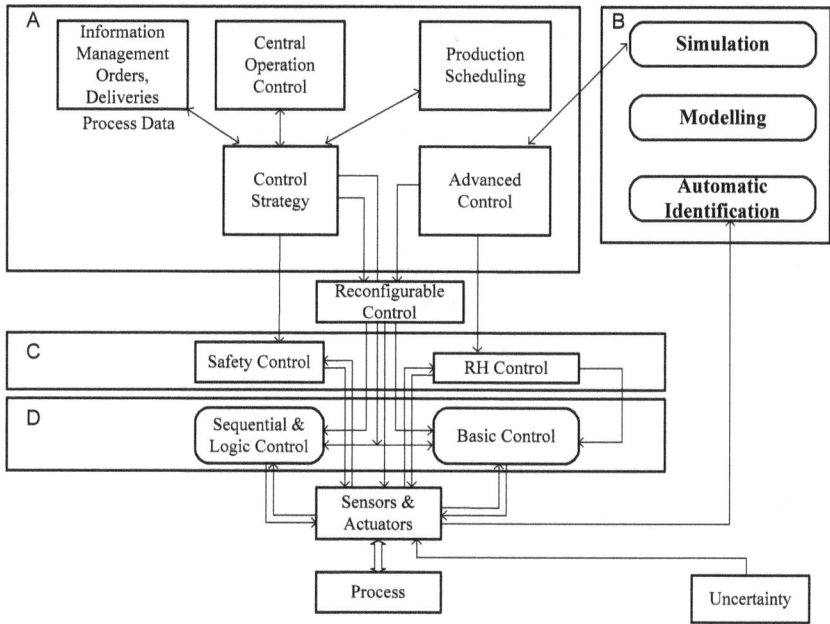

Fig. 5.14. Control and safety integrated architecture.

The four main holons are, from the upper level: D - classical control based on basic regulatory, sequential and logical, C - safety level based on safety instrumented systems (SIS) and the new paradigm RH control, B - remote level based on internet or cloud able to do automatic identification, modeling and simulations, A - management level, which has two main functions: management and supervisory control.

An architectural oriented approach used to develop advanced control integrated architecture open to incorporate more flexibility, must implement reconfigurable control to ensure the reaction to uncertainties.

5.7.3. Integrated Architecture for Real-time Control and Uncertainty Management

The purpose is the design of a real time control system for industrial installations of strategic importance, capable to do the on-line risk analysis and hazard prevention, incorporating the newest technologies, methods and methodologies in the field of control, system engineering

systems, hazard analysis, default tolerant control or complex process optimization. The architecture of the control system proposed is shown in Fig. 5.15. The process components are interconnected, the performances of one component depending partially of other subsystems, influences in the same time the behavior of the others.

The internal control parameters are represented by u, the measurable outputs of subsystems are represented by y, inputs from other components are represented by m and external outputs are represented by z. It is possible that for a specific subsystem some of the internal control parameters are identified with the inputs form different components and some of the external outputs be measurable internal outputs. The system, in its whole, functions under the influence of external perturbations, defaults or other uncertainties. The uncertainties are represented by f. At the lower level of the architecture we have the control loop, where r represents the reference and p is the control parameter.

At the supervisory level, three main functions are completed:

- The accommodation to default module which insures that the system maintains the dynamic performance under the influence of uncertainties, external perturbations and defaults;

- The optimization module, that insures the calculation of the optimum point of functioning in the presence of restrictions under the admissible range;

- The risk and hazard prevention analysis module insures the on-line risk analysis and intervenes for preventing the hazard and risk state.

The functions are done simultaneously but not independently, each module using the data received from the other modules. As shown, the accommodation to default module realizes the following functions: default detection, default isolation, default estimation and default tolerant control.

Based on the information received from the process, the residual value is calculated (command values and measured outputs for each subsystem). The residual value, nonlinear function of input/output is compared with a set point in order to identify if the situation is of default. The selection of the set point is extremely important; a too small value can lead to false alarms and a too high value leading to the lack in detection of defaults.

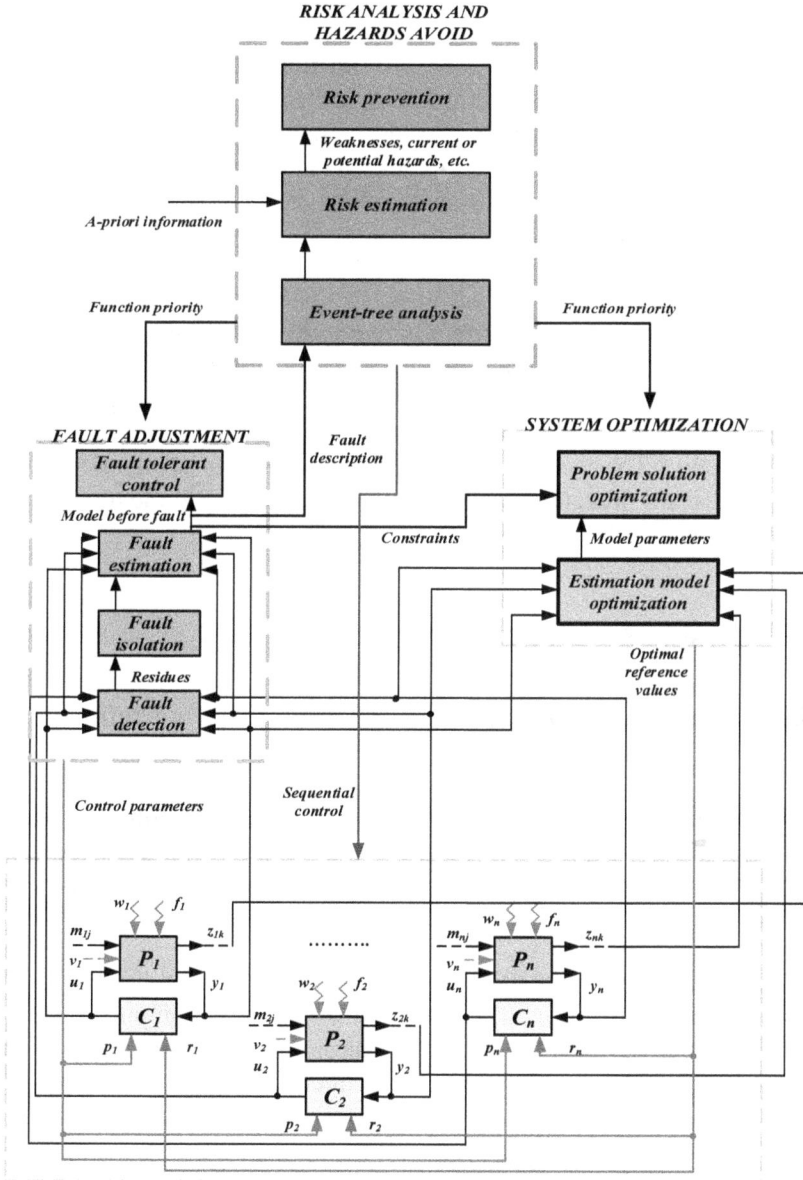

Fig. 5.15. System architecture.

If the module detects a default, the symptoms are sent to the isolation module that identifies the default components: sensors, actuators, technological equipment, and communications. The isolations contain two steps: the generation of generalized or structured residues and the comparison with the reference values associated to each set.

The optimization module realizes two functions: model estimation and problem solving. Based on chosen criteria: material or energy consumptions, quantity/time, product quality, etc., the process measurable variables can be established and also the corresponding values for the quality variables as inputs to the estimation module. The optimization module transmits the reference values for reaching the optimum point.

There is a strong interrelationship between the system optimization and default accommodation, because the emergence of a default influences the system behavior as a whole in the way that the functioning point can be modified most of the times by coming out of the range close to the nominal point. The accommodation module sends also information to the risk and hazard prevention analysis module, referring to the description of the detected default.

5.8. Results and Discussion

5.8.1. Evaluation of the Response Time for Uncertainty Control

Solutions can be conceived for reducing the uncertainty through causes and potential solutions identification and through supplementary functionalities of the system, in order to have a more performing response. This approach presumes the implementation of the following:

- The uncertainty control through management of request. Even if this is a recent approach, the management of the request is compulsory to be included in any system at the stage of design, through this insuring not only the connection to the market request variations, but also the maintenance of the acceptable performances in uncertainty conditions.

- Passive protection through robustness implementation. The classical method based on redundancy must also be completed with the implementation of the regulations of safety instrumented systems. The robustness, seen as the ability of the system to preserve the operational capabilities under different circumstances, can be reached through other methods.

- Active protection through flexibility and adaptability enhancement. There are approaches that, in order to maintain the functionality of the system, insure from the design stage, the degraded functioning of the system and of the process in uncertainty conditions with the maintenance of the minimal performances.

5.8.2. Methods for Response Elaboration

There are very few works for the development of theoretical methods that allow the utilization of algorithms for the response elaboration, capable of insuring the functioning in uncertain conditions. We can highlight the launching of the Risk and Hazard Control paradigm for the subsequent propositions also: architecture, strategy, computational modules. For implementing the tactical and strategic decisions, the most used approach is that of experience learning. This method can be passive or active. It is based on the recording of information, analysis and deduction of potential evolutions and tendencies of the uncertain situation. Based on this, the action options are estimated and strategies are implemented. Another approach consists in combining two methods – that of the time scale and that of the response module, expanding the area of analysis, synthesis, planning and system implementation. The Table 5.1 synthesizes the possible actions at every stage that leads to modification in the design of the process control system in order to insure the robustness and flexibility.

Passive or active system modifications can be done from the design stage if they can be seen as systems characteristics through hardware, software and communication methods. The mode in which these can be fulfilled by the working system derives from the influence of each attribute on those, as seen in Table 5.2.

From Table 5.3 also results which of these attributes can be implemented in diverse stages of the system's life cycle: design, execution, commissioning, operation, maintenance, and upgrade.

Table 5.1. Possible actions to achieve implementation level

Time scale and response mode	Management uncertainty	System modification	
		Passive: Robustness	Active: Flexibility
Operational	Correction of new sources of variation highlighted by the statistical analysis	Adapting of the execution speed	Implementing modules and functional blocks
Tactical	Investment in new measurements and in precise evaluation methods	Robust design selecting processing periods that cover diverse variations	Organizing the technological area as a multi agent holarchy
Strategic	Implementing of a working system with the clients that insures better quality and lower cost	Elaboration of a technology capable to respond to the precision and quality demands on the long term	Creation of a concurrent suppliers network for continuous selection

Table 5.2. Methods to achieve attributes

	Modify	Evolution	Reconfiguration	Adaptability	Sustainability	Modularity	Standardization	Stability	Wide area	Standard interfaces	Plug and Play	Redundancy	Holonic structure
Robustness	□	□	□	□		□	□	□	□		□		
Flexibility	□	□	□	□			□	□		□	□	□		□

Table 5.3. Development stages versus system attributes

	Modify	Evolution	Reconfiguration	Adaptability	Sustainability	Modularity	Standardization	Stability	Wide area	Standard interfaces	Plug and Play	Redundancy	Holonic structure
Projection	□	□	□	□	□	□	□	□	□	□	□	□	□
Execution	□	□				□	□	□	□	□	□	□	
Startup						□	□	□	□	□		□	
Operation	□	□	□	□	□		□	□					□
Maintenance	□		□	□	□			□	□	□	□	□	
Up-grade	□	□	□	□	□			□	□			□	□

5.8.3. Developed Platforms and Case Studies

The totally new approach in process control system engineering, based on new algorithms, scalable and modular architectures and platforms, RH control is industry independent. The capability of the systems to model and implement the 4 states, start-up, nominal, RH, shut-down, having 4 different strategies and the capability to change the state according to the functional parameters can be taken in consideration by CE. The diagnosis system, hosted remotely, will be continuously improved by gathering knowledge from various applications, based on identified problems, the solutions offered and their impact on the plant performance. The correlation factor between these different applications will influence future decisions. This way, the required period of time for solving a problem will be minimized, as well as the time that a plant needs to be shut down because of the instrumentation process control strategy.

Some of the expected results are an integrated exploitation of a collection of heterogeneous technologies for the prevention of atypical situations related to the safety of an industrial complex and determining the suitability of function blocks and OPC based development for integrated control systems construction.

From the user's point of view, the accomplishment is that RH Control will allow the integration of the preventive and corrective aspects of safety, which were dealt with, until this moment, in separate ways. Another advantage arises from being able to automatically take into account the constraints posed by the current plant situation and the ongoing maintenance operations.

The results achieved so far within the R&D project "Help Center and platform for remote diagnosis and remote intervention for the management of plants in hazardous situations – PH Center" [18] will be used to develop and implement the hierarchically superior level for safety and security problems. The work carried out in the project establishes the baselines for a new architecture of process control taking into consideration the remote operation.

At present, more than 10 applications are installed and under operation after a period of user validation and evaluation:

- Simulator for a simple process - controlling the level of liquid in a tank – Fig. 5.16;

- Building Management System (BMS) designed for a supermarket – Fig. 5.17;

- LPG process control;

- Burner Management System for a thermopower plant;

- Gateway and protocol converter for Wind Power Generators;

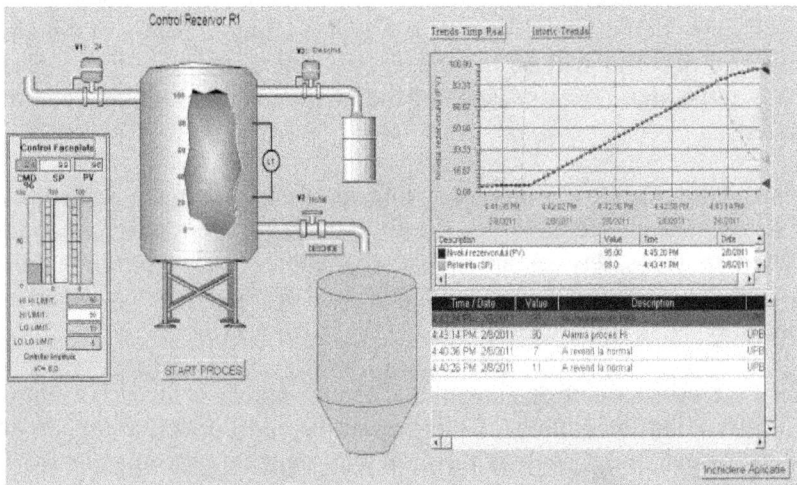

Fig. 5.16. The simulated process.

Also, we are currently working on including new connections to the PH Center, two SCADA systems (Siemens PLC, Seneca controllers, PC Vue software) from Water distribution networks (Fig. 5.18).

The generic products constructed within the project are truly reusable and can be exploitable components of other implementations.

The meantime, the results achieved underlay the feasibility of the idea. This statement is based on two reasons:

Demo-applications have been designed according to real plant requirements with a large involvement of plant staff.

Supermarket

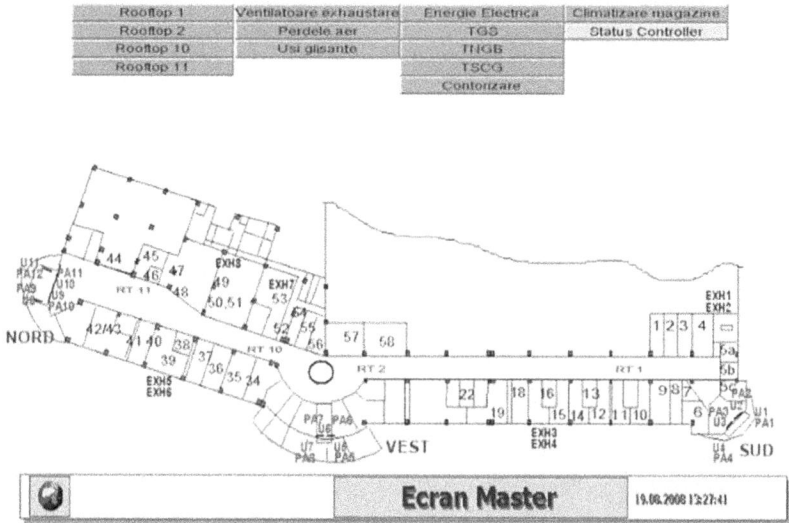

Fig. 5.17. Supermarket BMS – remote connection main screen.

5.9. Conclusions

The modern goal of process control is to keep the functionality of the process even at levels of less performance. The proposed solution focuses on implementing risk and hazard control paradigm. The approach proposed in this chapter aims to integrate the emerging technology: Risk and Hazard Control as control strategies for uncertainty management. Control Strategy is a planned set of controls, checks and sequences, derived from current product and process, having the target to assure process performance and product quality.

Extension of V method to design control systems for uncertain situation and combining the two methods – that of the time scale and that of the response module are the main results of research. Robustness and flexibility attributes of the system ensure the target of performing in uncertain conditions. Design meets the intended needs of customers, the quality goals to eliminate nonconformance and minimizes variation around appropriate targets, at minimum cost. Further work will be focused on the implementation of RH Control modules, using an aggregate of intelligent technologies: SOA approach based on Web Services, cooperative and consensual DSS and Multi-Agent System paradigm, with Holonic Multi-Agent System architecture.

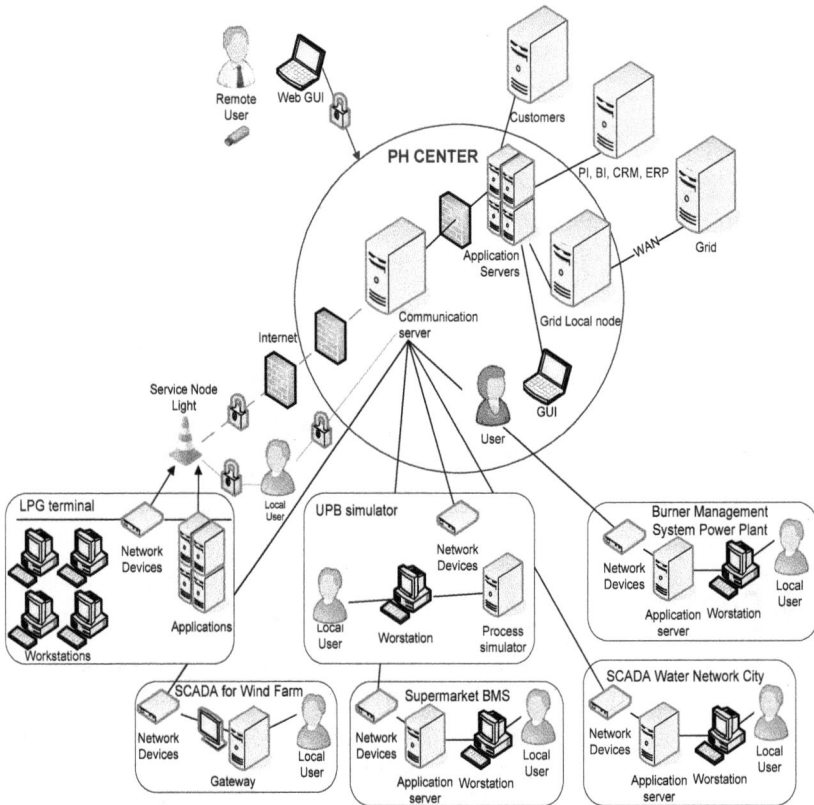

Fig. 5.18. PH Center connected to local control rooms.

Acknowledgement

This work was supported by the Romanian National Research Programme PNII, project: Cloud Architecture for an open Library of Complex re-Usable Logical function blocks for Optimized Systems – CALCULOS. and PN III, project: Solution for efficient integration of wind and solar generators in National Power Grid- INNOVATION.

References

[1]. American Institute of Chemical Engineers. Guidelines for Safe and Reliable Instrumented Protective Systems, 2007.

[2]. Asset Performance Management Supported by Reliability Engineering, *ReliaSoft*, Vol. 11, Issue 1.

[3]. Asish Ghosh, Dave Woll, Business Issues Driving Safety System Integration, *ARC Advisory Group*, 2006, pp. 1-12.

[4]. Ged Farnaby, Protect the plant. Leading edge trends in process control safety, *InTech*,2005.

[5]. Wes Iversen, The Great Safety Debate, *Automation World*, 2007.

[6]. David Hatch, Todd Stauffer, Operators on alert. Alarm standards, protection layers, HMI keys to keep plants safe, *InTech*, 2009.

[7]. Merry Spooner, Trevor MacDougall, Safety Instrumented Systems: can they be integrated but separate?, *DeltaV*, 2005.

[8]. Luiza Ocheană, Dan Popescu, Gheorghe Florea, Integrating versus Interfacing Safety and Security with Process Control System, in *Proceedings of the International Conference on Control Systems and Computer Science (CSCS'19)*, 2013, pp. 441-447.

[9]. Gary Stoneburner, Alice Goguen, Alexis Feringa, Risk Management Guide for Information Technology Systems. Recommendations of the National Institute for Standards and Technology, *NIST*, U.S.A., 2002.

[10]. Gheorghe Florea, Luiza Ocheana, RH Control, the next level of decision and intervention, in *Proceedings of the REV2011 - Remote Engineering & Virtual Instrumentation*, 2011, pp. 289-290.

[11]. Guttman Michael, Jason R. Matthews, The Object Technology Revolution, *John Wiley & Sons Inc.*, New York, NY, USA, 1995.

[12]. Stefan-Helmut Leitner, Wolfgang Mahnke, OPC UA – Service –oriented Architecture for Industrial Applications, *ABB Corporate Research Center*, Ladenburg, Germany, 2006.

[13]. Gheorghe Florea, Luiza Ocheană, Dan Popescu, Oana Rohat, Emerging Technologies – The Base For The Next Goal of Process Control – Risk and Hazard Control, in *Proceedings of the 11th WSEAS International Conference on Systems Theory and Scientific Computation (ISTASC'11)*, Firenze,2011, pp. 227-232.

[14]. National Research Council, Theoretical Foundations for Decision Making in Engineering Design, 2001. http://www.nap.edu/openbook.php?record_id=10566&page=8.

[15]. Daniel Merezeanu, Alexandra Ioana Florea (Ionescu), A Framework for Developing Lifecycle Management Based on IoT and RFID, *Journal of Control Engineering and Applied Infromatics*, Vol. 19, No. 1, 2017.

[16]. A. Gunasekaran, Concurrent engineering: a competitive strategy for process industries, *Journal of the Operational Research Society*, 1998, pp. 758-765.

[17]. Gheorghe Florea, Luiza Ocheana, Concurrent Engineering used to Implement Risk & Hazard Control, in *Proceedings of the 3rd International Conference on Advances in System Testing and Validation Lifecycle (VALID'11)*, 2011.

[18]. Luiza Ocheană, Dan Popescu, Gheorghe Florea, Risk And Hazard Prevention Using Remote Intervention, *U.P.B. Sci. Bull.*, Series C, Vol. 74, Issue 3, 2012.

[19]. Larsson T. & Skogestad S., Plantwide control - A review and a new design procedure, *Modeling, Identification and Control*, Vol. 21, No. 4, 2000.

[20]. Florea G. & Dobrescu R., Risk and Hazard Control the new process control paradigm, *Communications, Circuits and Educational Technologies*, 2014, pp. 141-149.

[21]. Levis A., System Architectures, Handbook of Systems Engineering and Management, *John Wiley & Sons*, 1999.

6.

A Slow-growing Hierarchy of Time-bounded Programs

Emanuele Covino and Giovanni Pani

6.1. Introduction

A complexity class is usually defined by setting an explicit bound on time and/or space resources used by a Turing Machine, or another equivalent model, during its computations. On the other hand, different approaches use logic and formal methods to provide languages for complexity-bounded computations; they study computational complexity without referring to external measuring conditions or to a particular machine model, but only by considering language restrictions or logical/computational principles implying complexity properties. In particular, complexity classes are characterized by means of recursive operators, with explicit syntactical restrictions on the role of variables. In this paper, we extend the result introduced in [1] by defining a resource-free characterization of register machines computing their output within polynomial time $O(n^k)$ and exponential $O(n^{n^k})$, for each finite k; we achieve this result by means of our version of *predicative recursion*, a new *diagonalization* operator, and a related programming language.

One of the first characterizations of the polynomial-time computable functions was given by Cobham [2], in which these functions are exactly those generated by bounded recursion on notation. The first predicative definitions of recursion can be found in the work of Simmons [3], Bellantoni and Cook [4], and Leivant [5-6]: they introduced a ramification principle that doesn't require that explicit bounds are imposed on the definition of functions, and they proved that this principle captures the class PTIMEF. Following this approach, several other

Emanuele Covino
Dipartimento di Informatica, Università di Bari, Italy

complexity classes have been characterized by means of unbounded and predicative operators: see, for instance, Leivant and Marion [7] and Oitavem [8] for PSPACEF and the class of the elementary functions; Clote [9] for the definition of a time/space hierarchy between PTIMEF and PSPACEF; Leivant [5, 10, 6] for a theoretical insight. All these approaches have been dubbed Implicit Computational Complexity: they share the idea that no explicitly bounded schemes are needed to characterize classes of functions and that, in order to do this, it suffices to distinguish between safe and normal variables (or, following [3], between dormant and normal ones) in the recursion schemes. Roughly speaking, the normal positions are used only for recursion, while the safe positions are used only for substitution. The two main objectives of this area are to find natural implicit characterizations of various complexity classes, thereby illuminating their nature and importance, and to design methods suitable for static verification of program complexity.

Our version of the *safe recursion* scheme on a binary word algebra is such that $f(x,y,za)=h(f(x,y,z),y,za)$; throughout this paper we will call x,y and z the auxiliary variable, the parameter, and the principal variable of a program defined by recursion, respectively. We don't allow the renaming of variable z as x, implying that the previous step program h cannot assign the value $f(x,y,z)$ of the being-defined program f to the principal variable z: in other words, we always know in advance the number of recursive calls of the step program in a recursive definition. We obtain that z is a *dormant* variable, according to Simmons [3], or a *safe* one, following Bellantoni and Cook [4]. Starting from a natural definition of constructors and destructors over an algebra of lists, we define the hierarchy of classes of programs T_k, with $k \geq 0$: programs in T_1 can be computed by register machines within linear time, and programs in T_{k+1} are obtained by one application of safe recursion to elements in T_k; we prove that programs defined in T_k are exactly those computable within time $O(n^k)$.

Using the definition of structured ordinals as given in [11], we introduce an operator of constructive *diagonalization*, and we extend the previously defined hierarchy to T_λ, with $\omega \leq \lambda \leq \omega^\omega$. Programs in T_{a+1} are obtained by one application of safe recursion to elements in T_a; if λ is a limit ordinal, and $\lambda_1, \ldots, \lambda_k, \ldots$ is the associated fundamental sequence, programs in T_λ are obtained by diagonalization on the previously defined sequence of classes $T_{\lambda_1}, \ldots, T_{\lambda_k}, \ldots$. This allows us to harmonize in a single hierarchy of classes of programs all the register machines with computing time bounded by polynomial time $O(n^k)$ and exponential

$O(n^{n^k})$, for each finite k. Similar results, achieved with different approaches, can be found in [11, 13-15].

Then, we restrict T_k to the hierarchy S_k, whose elements are the programs computable by a register machine in linear space. By means of a restricted form of composition between programs, we define a polytime-space hierarchy TS_{qp}, such that each program in TS_{qp} can be computed by a register machine within time $O(n^p)$ and space $O(n^q)$, simultaneously. Similar results can be found in [16] and [7], and they represent a preliminary step for an implicit classification of the hierarchy of time-space classes between PTIMEF and PSPACEF, as defined in [9].

The paper is organized as follows: in Section 2, we introduce the basic instructions of our language, the notion of composition of programs, and the classes of programs T_0 and T_1; in Section 3, we recall the definition of register machine, the model of computation underlying our characterization, and we prove that programs in T_1 capture exactly the computations performed by a register machine within linear time; in Section 4, we define the finite hierarchy T_k, with $k \geq 1$, and we prove that programs in this hierarchy capture the computations performed within polynomial time; in Section 5, we introduce the diagonalization operator, and we extend the finite hierarchy in order to capture the computations with time-complexity up to exponential time; in Section 6, we restrict the definition of composition, and we give a characterization of classes of programs that operates with a simultaneous polynomial bound on time and space complexity.

6.2. Basic Instructions and Definition Schemes

In this section, we introduce the basic operators of our programming language and the first two classes on which our hierarchy is based. The language is defined over a binary word algebra, with the restriction that words are packed into lists, with the symbol @ acting as a separator between each word. This allow us to handle a sequence of words as a single object. The basic instructions allow us to manipulate lists of words, with some restrictions on the renaming of variables; the language is completed by adding the definitions of recursion and of composition of programs.

153

6.2.1. Recursion-free Programs and Class T_0

B is the binary alphabet $\{0,1\}$. a, b, a_1, ... denote elements of B, and U, V, ..., Y denote words over B. p, q, ..., s, ... stand for lists in the form $Y_1@Y_2@...@Y_{n-1}@Y_n$. The ith *component* $(s)_i$ of $s = Y_1@Y_2@...@Y_{n-1}@Y_n$ is Y_i. $|s|$ is the length of the list s, that is the number of letters occurring in s. We write x, y, and z for the variables used in a program, and we write u for one among x, y, z. Programs are denoted with the letters f, g, h, and we write $f(x,y,z)$ for the application of the program f to variables x, y, z; some among the variables may be absent.

Definition 6.2.1: The *basic instructions* are:

1) The *identity* $I(u)$, that returns the value s assigned to u;

2) The *constructors* $C^a_i(s)$, that add the digit a at the right of the last digit of $(s)_i$, with $a = 0,1$ and $i \geq 1$;

3) The *destructors* $D_i(s)$, that erase the rightmost digit of $(s)_i$, with $i \geq 1$.

Constructors $C^a_i(s)$ and destructors $D_i(s)$ leave the input s unchanged if s has less than i components.

Example 6.2.1: Given the word $s=01@11@@00$, we have that $|s|=9$, and $(s)_2=11$. We also have $C^1_1(01@11)=011@11$, $D_2(0@0@)=0@@$, $D_2(0@@)=0@@$.

Definition 6.2.2: Given the programs g and h, f is defined by *simple schemes* if it is obtained by:

1) *Renaming* of x as y in g, that is, f is the result of the substitution of the value of y to all occurrences of x into g. Notation: $f = RNM_{x/y}(g)$;

2) *Renaming* of z as y in g, that is, f is the result of the substitution of the value of y to all occurrences of z into g. Notation: $f = RNM_{z/y}(g)$;

3) *Selection* in g and h, if for all s, t, r:

$$f(s,t,r) = \begin{cases} g(s,t,r) & \text{if the rightmost digit of } (s)_i \text{ is } b \\ h(s,t,r) & \text{otherwise} \end{cases}$$

with $i \geq 1$ and $b=0,1$. Notation: $f = \text{SEL}^b_i(g,h)$. The simple schemes are denoted with SIMPLE.

Example 6.2.2: if f is defined by $\text{RNM}_{xy}(g)$ we have that $f(t,r)=g(t,t,r)$. Similarly, f defined by $\text{RNM}_{zy}(g)$ implies that $f(s,t)=g(s,t,t)$. Let s be the word $00@1010$, and let f be defined as $\text{SEL}^0_2(g,h)$; we have that $f(s,t,r)=g(s,t,r)$, since the rightmost digit of $(s)_2$ is 0.

Definition 6.2.3: Given the programs g and h, f is the *safe composition* of h and g in the variable u if f is obtained by the substitution of h to u in g, if $u=x$ or $u=y$; the variable x must be absent in h, if $u=z$. Notation: $f = \text{SCMP}_u(h,g)$.

The reason of this particular form of composition of programs will be clear in the following section, where we will show to the reader how to combine composition and recursion in order to define new time-bounded programs.

Definition 6.2.4: A *modifier* is the safe composition of a sequence of constructors and a sequence of destructors.

Definition 6.2.5: T_0 is the class of programs obtained by closure of modifiers under selection and safe composition. Notation: $T_0 = (modifier; \text{SEL}, \text{SCMP})$.

Programs in T_0 modify their inputs according to the result of some test performed over a fixed number of digits.

6.2.2. Safe Recursion and Class T_1

In what follows we introduce the definition of our form of recursion and iteration *on notation* (see [4] and [10]).

Definition 6.2.6: Given the programs $g(x,y)$ and $h(x,y,z)$, $f(x,y,z)$ is obtained by *safe recursion* in the *basis* g and in the *step* h if for all s,t,r:

$$\begin{cases} f(s,t,a) & = & g(s,t) \\ f(s,t,ra) & = & h(f(s,t,r),t,ra) \end{cases}$$

with $a \in \boldsymbol{B}$. Notation: $f = \text{SREC}(g,h)$. In particular, $f(x,z)$ is defined by *iteration* of $h(x)$ if for all s,r:

$$\begin{cases} f(s,a) &= s \\ f(s,ra) &= h(f(s,r)) \end{cases}$$

with $a \in \boldsymbol{B}$. *Notation:* $f = \text{ITER}(h)$. *We write* $h^{|r|}(s)$ *for* $\text{ITER}(h)(s,r)$ *(i.e., the* $|r|^{th}$ *iteration of h on s).*

Definition 6.2.7: T_1 is the class defined by closure under safe composition and simple schemes of programs in T_0 and programs obtained by one application of ITER to T_0. Notation: $T_1 = (T_0, \text{ITER}(T_0); \text{SCMP, SIMPLE})$.

As we have already stated in the Introduction, we call x, y and z the auxiliary variable, the parameter, and the principal variable of a program obtained by means of the previous recursion scheme. Note that the renaming of z as x is not allowed (see Definition 6.2.2), and if the step program of a recursion is defined itself by means of safe composition of programs p and q, no variable x (i.e., no potential recursive calls) can occur in the program p, when p is substituted into the principal variable z of q (see Definition 6.2.3). These two restrictions implies that the step program of a recursive definition never assigns the recursive call to the principal variable. This is the key of the polynomial-time complexity bound intrinsic to our programs.

Definition 6.2.8: Given $f \in T_1$, the *number of components* of f is $max\{ i \mid D_i$ or $C^a{}_i$ or $\text{SEL}_i{}^b$ occurs in $f\}$. Notation: $\#(f)$. Given a program f, its *length* is the number of constructors, destructors and defining schemes occurring in the definition of f. Notation: $lh(f)$.

6.3. Computation by Register Machines

In this section, we recall the definition of register machine as presented in [6], and we give the definition of computation within a given time, space, or simultaneous bound. We prove that programs in T_1 are exactly those computable within linear time.

Definition 6.3.1: Given a free algebra A generated from constructors c_1, ..., c_n (with $arity(c_i) = r_i$), a *register machine* over A is a computational device M having the following components:

1) A finite set of *states* $S = \{s_0, \ldots, s_n\}$;

2) A finite set of *registers* $\Phi = \{\pi_0, \ldots, \pi_m\}$;

3) A collection of *commands*, where a command may be

a **branching** $s_i\pi_j s_{i_1}, \ldots, s_{i_k}$, such that when M is in the state s_i, switches to state $s_{i_1}, \ldots, s_{i_k}$ according to whether the main constructor (i.e., the leftmost) of the term stored in register π_j is $c_1, \ldots, c_k$;

a **constructor** $s_i\pi_{j_1}\ldots\pi_{j_{ri}}c_i\pi_l\, s_r$, such that when M is in the state s_i, stores in π_l the result of the application of the constructor c_i to the values stored in $\pi_{j_1}\ldots\pi_{j_{ri}}$, and switches to s_r;

a **p-destructor** $s_i\pi_j\pi_l s_r$ $(p \le max(r_i)_{i=1\ldots k})$, such that when M is in the state s_i, stores in π_l the pth subterm of the term in π_j, if it exists; otherwise, stores the term in π_j. Then it switches to s_r.

A *configuration* of M is a pair (s,F), where $s \in$ and $F: \Phi \to A$. M induces a transition relation $\equiv_M$ on configurations, where $\kappa \equiv_M \kappa'$ holds if there is a command of M whose execution converts the configuration κ in κ'. A *computation* of M on input $X_1, \ldots, X_p$ with output $Y_1, \ldots, Y_q$ is a sequence of configurations, starting with (s_0,F_0), and ending with (s_1,F_1) such that:

1) $F_0(\pi_{j'(i)}) = X_i$, for $1 \le i \le p$ and j' a permutation of the p registers;

2) $F_1(\pi_{j''(i)}) = Y_i$, for $1 \le i \le q$ and j'' a permutation of the q registers;

3) Each configuration is related to its successor by $\equiv_M$;

4) The last configuration has no successor by $\equiv_M$.

Definition 6.3.2: A register machine M *computes* the program f if, for all s,t,r, we have that $f(s,t,r)=q$ implies that M computes $(q)_1,\ldots,(q)_{\#(f)}$ on input $(s)_1,\ldots,(s)_{\#(f)}, (t)_1,\ldots,(t)_{\#(f)}, (r)_1,\ldots,(r)_{\#(f)}$.

Definition 6.3.3: Given a register machine M and the polynomials $p(n)$ and $q(n)$, for each input of overall length n,

1) M computes its output within time $O(p(n))$ if its computation runs through $O(p(n))$ configurations;

2) M computes its output in space $O(q(n))$ if, during the computation, the total length of the content of its registers is $O(q(n))$.

3) M computes its output with time $O(p(n))$ and space $O(q(n))$ if the two bounds occur simultaneously, during the same computation.

Note that the number of registers needed by M to compute a program f has to be fixed a priori (otherwise, we should have to define a family of register machines for each program to be computed, with each element of the family associated to an input of a given length). According to Definitions 6.2.8 and 6.3.2, M uses a number of registers which linearly depends on the highest component's index that f can manipulate or access with one among its constructors, destructors or selections; and which depends on the number of times a variable is used by f, that is, on the total number of different copies of the registers that M needs during the computation. Both these numbers are constant values, and can be detected before the computation occurs.

Unlike the usual operators *cons*, *head* and *tail* over Lisp-like lists, our constructors and destructors can have direct access to any component of a list, according to Definition 6.2.1. Hence, their computation by means of a register machine requires constant time, but it requires an amount of time which is linear in the length of the input, when performed by a Turing machine.

Codes. We write $s_i@F_j(\pi_0)@...@F_j(\pi_k)$ for the word that encodes a configuration (s_i, F_j) of M; each component is a binary word over $\{0,1\}$.

Lemma 6.3.1: f belongs to T_1 if and only if f is computable by a register machine within time $O(n)$.

Proof: To prove the first implication we show (by induction on the structure of f) that each $f \in T_1$ can be computed by a register machine M_f in time cn, where c is a constant which depends on the definition of f, and n is the length of the input.

Base. $f \in T_0$. This implies that f is obtained by closure of a number of modifiers under selection and safe composition; each modifier g can be computed within time bounded by $lh(g)$, the overall number of basic instructions in the definition of g, i.e., by a machine running over a constant number of configurations; the result holds, since the selection can be simulated by a branching, and the safe composition can be simulated by a sequence of register machines, one for each modifier.

Step. Case 1. $f = \text{ITER}(g)$, with $g \in T_0$. We have that $f(s,r)=g^{|r|}(s)$. A register machine M_f can be defined as follows: $(s)_i$ is stored in the register π_i $(i=1\ldots\#(f))$ and r is stored in the register $\pi_{\#(f)+1}$. M_f calls M_g (whose computation time is bounded by $lh(g)$) for $|r|$ times, using the appropriate destructor on register $\pi_{\#(f)+1}$; the computation stops, returning the final result, when $\pi_{\#(f)+1}$ is empty. Thus, M_f computes $f(s,r)$ within time $|r|lh(g)$.

Case 2. Let f be defined by simple schemes or safe composition of functions in T_0 or T_1. The result follows by direct simulation of the schemes.

In order to prove the second implication, we show that the behaviour of a k-register machine M, which operates in time cn can be simulated by a program in T_1. Let nxt_M be a program in T_0, that operates on a generic input $s=s_i@F_j(\pi_0)@\ldots@F_j(\pi_k)$ and that simulates a single transition of M; nxt_M checks the state and the content of the registers, and, accordingly, it updates the code of the state and the code of one among the registers by means of modifiers, depending on the definition of M. By means of c-1 safe compositions, we define nxt^c_M in T_0, which applies nxt_M to the word that encodes a configuration of M for c times. We define

$$\begin{cases} linsim_M(x,a) & =x \\ linsim_M(x,za) & =nxt^c_M(linsim_M(x,z)) \end{cases}$$

$linsim_M(s,r)$ belongs to T_1 and iterates $nxt_M(s)$ for $c|r|$ times, returning the code of the configuration that contains the final result of M.

6.4. The Time Hierarchy

In this section, we recall the definition of the classes of programs T_0 and T_1; we define our hierarchy of classes of programs T_k, with $k \geq 0$, and we prove the relation with the classes of register machines which compute their output within a polynomially-bounded amount of time.

Definition 6.4.1:

1) $\text{ITER}(T_0)$ denotes the class of programs obtained by one application of iteration to programs in T_0;

2) T_1 is the class of programs obtained by closure under safe composition and simple schemes of programs in T_0 and programs in ITER(T_0). Notation: $T_1 = (T_0, \text{ITER}(T_0); \text{SCMP, SIMPLE})$;

3) T_{k+1} is the class of programs obtained by closure under safe composition and simple schemes of programs in T_k and programs in SREC(T_k), with $k \geq 1$. Notation: $T_{k+1} = (T_k, \text{SREC}(T_k); \text{SCMP, SIMPLE})$.

Lemma 6.4.1: Each $f(s,t,r)$ in T_k can be computed by a register machine within time bounded by $|s| + lh(f)(|t| + |r|)^k$, with $k \geq 1$.

Proof: Base. $f \in T_1$. The considerable case is when f is in the form ITER(h), with $h \in T_0$. In Lemma 6.3.1 (step, case 1) we have shown that $f(s,r)$ can be computed within time $|r| lh(h)$.

Step. $f \in T_{p+1}$. The most significant case is when $f = \text{SREC}(g,h)$. By the inductive hypothesis there exist two register machines M_g and M_h which compute g and h within the required time. Let r be the word $a_1...a_{|r|}$; recalling that $f(s,t,ra) = h(f(s,t,r),t,ra)$, we define a register machine M_f that calls M_g on input s,t, and calls M_h for $|r|$ times on input stored into the appropriate set of registers (in particular, the result of the previous recursive step has to be stored always in the same register). By inductive hypothesis, M_g needs time $|s| + lh(g)(|t|)^p$ in order to compute g; for the first call of the step program h, M_h needs time $|g(s,t)| + lh(h)(|t| + |a_{|r|}-1 a_{|r|}|)^p$. After $|r|$ calls of M_h, the final configuration is obtained within overall time $|s| + max(lh(g), lh(h))(|t| + |r|)^{p+1} \leq |s| + lh(f)(|t| + |r|)^{p+1}$.

Lemma 6.4.2: The behaviour of a register machine which computes its output within time $O(n^k)$ can be simulated by a program f defined in T_k, with $k \geq 1$.

Proof: Let M be a register machine respecting the hypothesis. As we have already seen, there exists $nxt_M \in T_0$ such that, for input the code of a configuration of M, it returns the code of the configuration induced by the relation $\equiv_M$. Given a fixed i, we define as follows the program σ_i by means of i safe recursions nested over nxt_M, such that it iterates nxt_M on input s for n^i times, with n the length of the input:

$\sigma_0 = \text{ITER}(nxt_M)$ and $\sigma_{n+1} = \text{RMN}_{z/y}(\gamma_{n+1})$, where $\gamma_{n+1} = \text{SREC}(\sigma_n, \sigma_n)$.

We have that $\sigma_o(s,t) = nxt_M^{|t|}(s)$, $\sigma_{n+1}(s,t) = \gamma_{n+1}(s,t,t)$ and

$$\begin{cases} \gamma_{n+1}(s,t,a) & = & \sigma_n(s,t) \\ \gamma_{n+1}(s,t,ra) & = & \sigma_n(\gamma_{n+1}(s,t,r),t) \\ & = & \gamma_n(\gamma_{n+1}(s,t,r),t,t) \end{cases}$$

In particular, we have

$\sigma_1(s,t) = \gamma_1(s,t,t) = \sigma_0(\sigma_0(\dots\sigma_0(s,t)\dots)) = nxt_M^{|t|^2}(s)$ (σ_0 is repeated $|t|$ times) and

$\sigma_2(s,t) = \gamma_2(s,t,t) = \sigma_1(\sigma_1(\dots\sigma_1(s,t)\dots)) = nxt_M^{|t|^3}(s)$ (σ_1 is repeated $|t|$ times).

By induction, we see that σ_{k-1} iterates nxt_M on input s for $|t|^k$ times, and that it belongs to T_k. The result holds defining $f(t) = \sigma_{k-1}(t,t)$, with t the code of an initial configuration of M.

Theorem 6.4.1: A program f belongs to T_k if and only if f is computable by a register machine within time $O(n^k)$, with $k \geq 1$.

Proof: By Lemma 6.4.1 and Lemma 6.4.2.

We recall that register machines are polytime reducible to Turing machines; thus, the sequence of classes T_k captures PTIMEF (see [6] and [12] for similar characterization of this complexity class).

6.5. Extending the Polynomial-Time Hierarchy to Transfinite

In this section, we extend the definition of the classes of programs T_k, with $k \geq 1$, to a transfinite hierarchy of classes; in order to do this, we recall the definition of structured ordinals and of hierarchies of slow/fast growing functions, as reported in [11]. We introduce a natural slow growing function B, and we define the *diagonalization* at a limit ordinal λ, which is based on the sequence of classes $T_{\lambda_1}, \dots, T_{\lambda_n}, \dots$ associated with the fundamental sequence of λ. A similar constructive operator can be found in [15] and [12]. We prove that this transfinite hierarchy of programs characterizes the classes of register machines computing their output with a time-complexity bound between $O(n^k)$ and $O(n^{n^k})$ (with $k \geq 1$ and n the length of the input), that is, the computations between polynomial- and exponential-time.

161

6.5.1. Structured Ordinals and Hierarchies

Following [11], we denote limit ordinals with Greek small letters α, β, λ, ..., and we denote with λ_i the ith element of the fundamental sequence assigned to λ. For example, ω is the limit ordinal of the fundamental sequence 1, 2, ...; and ω^2 is the limit ordinal of the fundamental sequence ω, $\omega 2$, $\omega 3$, ..., with $(\omega^2)_k = \omega k$.

The slow-growing functions G_α: N→N are defined by the recursion

$$
\begin{cases}
G_0(n) & = & 0 \\
G_{\alpha+1}(n) & = & G_\alpha(n)+1 \\
G_\lambda(n) & = & G_{\lambda_n}(n)
\end{cases}
$$

The *fast-growing functions* F_α: N→N are defined by the recursion

$$
\begin{cases}
F_0(n) & = & n+1 \\
F_{\alpha+1}(n) & = & F_\alpha^{n+1}(n) \\
F_\lambda(n) & = & F_{\lambda_n}(n)
\end{cases}
$$

We define the *slow-growing functions* B_α: N→N by means of the recursion

$$
\begin{cases}
B_0(n) & = & 1 \\
B_{\alpha+1}(n) & = & nB_\alpha(n) \\
B_\lambda(n) & = & B_{\lambda_n}(n)
\end{cases}
$$

Note that $B_k(n)=n^k$, $B_\omega(n)=n^n$, $B_{\omega+k}(n)=n^{n+k}$, $B_{\omega k}(n)=n^{nk}$, $B_{\omega^k}(n)=n^{n^k}$, and $B_{\omega^\omega}(n)=n^{n^n}$; moreover, we have that $B_{\alpha+\beta}(n)=B_\alpha(n) \cdot B_\beta(n)$, and that $G_{\omega^\alpha}(n)=n^{G_\alpha(n)}=B_\alpha(n)$.

6.5.2. Diagonalization and Transfinite Hierarchy

The finite hierarchy T_0, T_1, T_2, ..., T_k, ... captures the register machines that compute their output with time in $O(1)$, $O(n)$, $O(n^2)$, ..., $O(n^k)$, ..., respectively. Jumping out of the hierarchy requires something more than safe recursion. A possible approach consist in defining a kind of ranking function that counts the number of nested recursion violating our "no-bad-renaming" rule or, in general, not respecting the predicative definition of a program. A class of time-bounded register machines is associated to each level of violation. This idea was introduced in [17].

On the other hand, given a limit ordinal λ, we introduce a new operator that *diagonalizes* at level λ over the classes T_{λ_i}, with $i \geq 0$, selecting programs in that hierarchy of classes according to the length of the input. There is no circularity in a program defined by diagonalization, and we believe that this program isn't less predicative than a program defined by safe recursion. For instance, at level ω, we are able to select programs in the sequence T_i, with $i \geq 0$, where the value of i depends on the length of the input; thus, this level of diagonalization captures the class of all register machines whose computation is bounded by a polynomial. By extending this approach to the next levels of structured ordinals, we are able to reach the machines computing within exponential time.

Definition 6.5.1: Given a limit ordinal λ with the fundamental sequence $\lambda_0, \ldots, \lambda_k, \ldots$, and given an enumerator program q such that $q(\lambda_i) = f_{\lambda_i}$, for each $i \geq 0$, the program $f(x,y)$ is defined by *diagonalization* at λ if for all s,t:

$$f(s,t) = \mathrm{ITER}^{|t|}(q(\lambda_{|t|}))(s,t),$$

where

$$\begin{cases} \mathrm{ITER}^1(p)(s,t) & = & \mathrm{ITER}(p)(s,t) \\ \mathrm{ITER}^{k+1}(p)(s,t) & = & \mathrm{ITER}(\mathrm{ITER}^k(p))(s,t) \end{cases}$$

and f_{λ_i} belongs to a previously defined class C_{λ_i}, for each i. Notation: $f = \mathrm{DIAG}(\lambda)$.

Note that the previous definition requires that $f_{\lambda_i} \in C_{\lambda_i}$, but no other requirements are made on how the C_{λ_i} classes are built. In what follows, we introduce the transfinite hierarchy of programs, with an important restriction on the definition of the C_{λ_i}.

Definition 6.5.2: Given $\lambda < \omega^\omega$, T_λ is the class of programs obtained by

1) Closure under safe composition and simple schemes of programs in T_α and programs in $\mathrm{SREC}(T_\alpha)$, if $\lambda = \alpha + 1$. Notation: $T_{\alpha+1} = (T_\alpha, \mathrm{SREC}(T_\alpha); \mathrm{SCMP}, \mathrm{SIMPLE})$;

2) Closure under simple schemes of programs obtained by a single application of diagonalization at λ, if λ is a limit ordinal, with $f_{\lambda_i} \in T_{\lambda_i}$, for each λ_i in the fundamental sequence of λ. Notation: $T_\lambda = (\mathrm{DIAG}(\lambda); \mathrm{SIMPLE})$.

Lemma 6.5.1: Each $f(s,t,r)$ in T_λ can be computed by a register machine within time $O(B_\lambda(n))$, with $\lambda < \omega^\omega$.

Proof: By induction on λ. We have three cases:

(1) $\lambda = k < \omega$; because $B_k(n) = n^k$, the result follows from Lemma 6.4.1.

(2) $\lambda = \beta+1$; this implies that $f \in T_{\beta+1}$, and the relevant subcase is when $f = \mathrm{SREC}(g,h)$, with both g and h belonging to T_β. By the inductive hypothesis, there exist the register machines M_g and M_h computing g and h, respectively, within time bounded by $B_\beta(n)$. A register machine M_f can be defined, such that it calls M_g on input s,t, and calls M_h for $|r|$ times on input stored into the appropriate set of registers. M_f needs time equal to $B_\beta(n) + |r|B_\beta(n)$ to perform this computation; thus, the overall time is bounded by $B_{\beta+1}(n)$, by definition of B.

(3) λ is a limit ordinal; this means that f is defined by $\mathrm{DIAG}(\lambda)$, that is $f(s,t) = \mathrm{ITER}^{|t|}(q(\lambda_{|t|}))(s,t)$, with λ_i the fundamental sequence of λ, and $q(\lambda_i) = f_{\lambda_i} \in T_{\lambda_i}$, with $i < \omega$. By induction on the length of the input, we have that $f(s,a) = \mathrm{ITER}^{|a|}(q(\lambda_{|a|}))(s,a) = s$; obviously, there exists a register machine computing the result within time $B_{\lambda_{|a|}}(n)$. As for the step case we have that

$$f(s,ta) = \mathrm{ITER}^{|ta|}(q(\lambda_{|ta|}))(s,ta) = \mathrm{ITER}(\mathrm{ITER}^{|t|}(q(\lambda_{|ta|}))(s,t);$$

by inductive hypothesis, there exist a sequence of register machines $M_{\lambda_{|ta|}}$ computing the programs $q(\lambda_{|ta|})$ within time $B_{\lambda_{|ta|}}(n)$. We define a register machine M_f such that, on input s, t, it iterates $|t|$ times $M_{\lambda_{|ta|}}$, within time $B_{\lambda_{|ta|}}(n) \leq B_\lambda(n)$.

Lemma 6.5.2: The behaviour of a register machine which computes its output within time $O(B_\lambda(n))$ can be simulated by a program f defined in T_λ.

Proof: Let M be a register machine respecting the hypothesis. As we have already seen, there exists a program $nxt_M \in T_0$ such that, for input the code of a configuration of M, it returns the code of the configuration induced by the relation $\equiv_M$. We have three cases:

(1) $\lambda = k < \omega$; the result follows from Lemma 6.4.2.

(2) $\lambda = \beta+1$; we define the program σ_λ as follows: $\sigma_{\beta+1}:=\mathrm{RNM}_{z/y}(\gamma_{\beta+1})$, where $\gamma_{\beta+1}:=\mathrm{SREC}(\sigma_\beta,\sigma_\beta)$. We have that $\sigma_{\beta+1}(s,t)=\gamma_{\beta+1}(s,t,t)$, and

$$
\begin{cases}
\gamma_{\beta+1}(s,t,a) & = & \sigma_\beta(s,t) \\
\gamma_{\beta+1}(s,t,ra) & = & \sigma_\beta(\gamma_{\beta+1}(s,t,r),t) \\
& = & \gamma_\beta(\gamma_{\beta+1}(s,t,r),t,t)
\end{cases}
$$

In particular, we have $\sigma_{\beta+1}(s,t) = \gamma_{\beta+1}(s,t,t) = \sigma_\beta(\sigma_\beta(\ldots\sigma_\beta(s,t)\ldots,t),t)$, $|t|$ times. By induction, σ_β iterates nxt_M on its input s for $B_\beta(|t|)$ times, and it belongs to T_β. The result holds observing that $\sigma_{\beta+1}$ iterates nxt_M for $|t|B_\beta(|t|) = B_{\beta+1}(|t|)$ times.

(3) λ is a limit ordinal. Let $\lambda_1, \ldots, \lambda_n, \ldots$ the fundamental sequence associated to λ, and $\sigma_{\lambda_1}, \ldots, \sigma_{\lambda_n}, \ldots$ the sequence of programs enumerated by g, such that $g(\lambda_i) = \sigma_{\lambda_i} \in T_{\lambda_i}$, with $i<\omega$. We define γ_λ by diagonalization at limit λ. With a fixed input s,t, we have that $\gamma_\lambda(s,t) = \mathrm{ITER}^{|t|}(g(\lambda_{|t|}))(s,t) = \mathrm{ITER}^{|t|}(\sigma_{\lambda_{|t|}})(s,t)$. The programs $g(\lambda_{|t|})$ are defined in $T_{\lambda_{|t|}}$, and they iterate the program nxt_M on its input for $B_{\lambda_{|t|}}(|t|)$ times; this implies that γ_λ iterates nxt_M for $B_{\lambda_{|t|}}(|t|) = B_\lambda(|t|)$, for each t.

Theorem 6.5.1: A program f belongs to T_α if and only if f is computable by a register machine within time $O(B_\alpha(n))$, with $\alpha < \omega^\omega$.

Proof: By Lemma 6.5.1 and Lemma 6.5.2.

6.6. The Time-Space Hierarchy

In this section, we introduce a restricted version of the previously defined time-hierarchy of recursive programs, and we prove the equivalence with the classes of register machines, computing their output with a simultaneous bound on time and space.

6.6.1. Recursion-free Programs and Class S_0

The reader should refer to Section 6.2 for the definitions of basic instructions (the *identity* $I(u)$, the *constructors* $C^a_i(s)$, and the *destructors* $D_i(s)$); simple schemes (the *renaming* $RNM_{x/y}(g)$ and $RMN_{z/y}(g)$, and the *selection* $SEL^b_i(g,h)$); and safe composition $SCMP_u(h,g)$. In particular, we recall that a *modifier* is obtained by the safe composition of a sequence of constructors and a sequence of destructors; according to Definition 6.2.5, the class T_0 is the class of programs defined by closure of modifiers under SEL and SCMP.

Definition 6.6.1: Given $f \in T_0$, the *rate of growth rog(f)* is such that

1) if f is a modifier, $rog(f)$ is the difference between the number of constructors and the number of destructors occurring in its definition;

2) if $f = SEL^b_i(g,h)$, then $rog(f)$ is $max(rog(g), rog(h))$;

3) if $f = SCMP_u(h,g)$, then $rog(f)$ is $max(rog(g), rog(h))$.

Definition 6.6.2: S_0 is the class of programs in T_0 with non-positive rate of growth, that is $S_0 = \{f \in T_0 \mid rog(f) \leq 0\}$.

Note that all programs in S_0 modify their inputs according to the result of some test performed over a fixed number of digits and, moreover, they cannot return values longer than their input.

6.6.2. Safe Recursion and Class S_1

As written in Section 2.1, a program $f(x,y,z)$ is defined by *safe recursion* in the *basis* $g(x,y)$ and in the *step* $h(x,y,z)$ if for all s,t,r:

$$\begin{cases} f(s,t,a) &= g(s,t) \\ f(s,t,ra) &= h(f(s,t,r),t,ra) \end{cases}$$

In this case, f is denoted with $SREC(g,h)$. In particular, $f(x,z)$ is defined by *iteration* of $h(x)$ if for all s,r:

$$\begin{cases} f(s,a) &= s \\ f(s,ra) &= h(f(s,r)) \end{cases}$$

In this case, f is denoted with ITER(h), and we write $h^{|r|}(s)$ for ITER(h)(s,r).

Definition 6.6.3:

1) ITER(S_0) denotes the class of programs obtained by one application of iteration to programs in S_0;

2) S_1 is the class of programs obtained by closure under safe composition and simple schemes of programs in S_0 and programs in ITER(S_0). Notation: $S_1 = (S_0, \text{ITER}(S_0); \text{SCMP}, \text{SIMPLE})$;

3) S_{k+1} is the class of programs obtained by closure under simple schemes of programs in S_k and programs in SREC(S_k). Notation: $S_{k+1} = (S_k, \text{SREC}(S_k); \text{SIMPLE})$.

Hence, hierarchy S_k, with $k \geq 0$, is a version of T_k in which each program returns a result whose length is *exactly* bounded by the length of the input; this does not happen if we allow the closure of S_k under SCMP. We will use this result to evaluate the space complexity of our programs.

Definition 6.6.4: Given the programs g and h, f is obtained by *weak composition* of h in g if $f(x,y,z) = g(h(x,y,z),y,z)$. Notation: $f = \text{WCMP}(h,g)$.

In the *weak* form of composition the program h can be substituted only in the variable x, while in the *safe* composition the substitution is possible in all variables.

Definition 6.6.5: For all $p,q \geq 1$, TS_{qp} is the class of programs obtained by weak composition of h in g, with $h \in T_q$, $g \in S_p$, and $q \leq p$.

Lemma 6.6.1: For all f in S_p, we have $|f(s,t,r)| \leq max(|s|,|t|,|r|)$.

Proof: By induction on p. Base. The relevant case is when $f \in S_1$ and f is defined by iteration of $g \in S_0$ (that is, $rog(g) \leq 0$). By induction on r, we have that $|f(s,a)| = |s|$, and $|f(s,ra)| = |g(f(s,r))| \leq |f(s,r)| \leq max(|s|,|r|)$.

Step. Given $f \in S_{p+1}$, defined by SREC in g and h in S_p, we have $|f(s,t,a)| = |g(s,t)| \leq max(|s|,|t|)$, by definition of f and by inductive hypothesis, and

$$|f(s,t,ra)| \; = \; |h(f(s,t,r),t,ra)|$$
$$\leq \; \max(|f(s,t,r)|,|t|,|ra|)$$
$$\leq \; \max(\max(|s|,|t|,|r|),|t|,|ra|)$$
$$\leq \; \max(|s|,|t|,|ra|)$$

by definition of f, inductive hypothesis on h and induction on r.

Lemma 6.6.2: Each f in TS_{qp} can be computed by a register machine within time $O(n^p)$ and space $O(n^q)$, with $1 \leq q \leq p$.

Proof: Let f be in TS_{qp}. By Definition 6.6.5, f is defined by weak composition of $h \in T_q$ into $g \in S_p$, that is, $f(s,t,r) = g(h(s,t,r),t,r)$. The Theorem 6.5.1 states that there exists a register machine M_h, which computes h within time n^q, and there exists another register machine M_g, which computes g within time n^p. Since g belongs to S_p, Lemma 6.6.1 holds for g; hence, the space needed by M_g is at most n. We define now a machine M_f that, by input s,t,r, performs the following steps: (1) it calls M_h on input s,t,r; (2) it calls M_g on input $h(s,t,r),t,r$, stored in the appropriate registers.

According to Lemma 6.4.2, M_h needs time equal to $|s|+lh(h)(|t|+|r|)^q$ to compute h, and M_g needs $|h(s,t,r)|+lh(g)(|t|+|r|)^p$ to compute g. This happens because Lemma 6.4.2 shows, in general, that the time used by a register machine to compute a program is bounded by a polynomial in the length of its inputs, but, more precisely, it shows that the time complexity is linear in $|s|$. Moreover, since in our language there isn't any renaming of x as z, M_f never moves the content of a register associated to $h(s,t,r)$ into another register and, in particular, into a register whose value plays the role of recursive variable. Thus, the overall time-bound is $|s|+lh(h)(|t|+|r|)^q + lh(g)(|t|+|r|)^p$, which can be reduced to n^p, being $q \leq p$. M_h requires space n^q to compute the value of h on input s,t,r; as we noted above, the space needed by M_g for the computation of g is linear in the length of the input, and thus the overall space needed by M_f is still $O(n^q)$.

Lemma 6.6.3: A register machine which computes its output within time $O(n^p)$ and space $O(n^q)$ can be simulated by a program f in TS_{qp}.

Proof: Let M be a register machine, whose computation is time-bounded by n^p and, simultaneously, is space-bounded by n^q. M can be simulated by the composition of two machines, M_h (time-bounded by n^q), and M_g

(time-bounded by n^p and, simultaneously, space-bounded by n): the former delimits (within n^q steps) the space that the latter will successively use in order to simulate M.

By Theorem 6.5.1 there exists $h \in T_q$ that simulates the behaviour of M_h, and there exists $g \in T_p$ that simulates the behaviour of M_g; this is done by means of nxt_g, which belongs to S_0, since it never adds a digit to the description of M_g without erasing another one. According to the proof of Lemma 6.4.1, we are able to define $\sigma_{n-1} \in S_n$, such that $\sigma_{n-1}(s,t) = nxt_g^{|t|^n}$. The result holds defining $sim(s) = \sigma_{p-1}(h(s),s) \in TS_{qp}$.

Theorem 6.6.1: f belongs to TS_{qp} if and only if f is computable by a register machine within time $O(n^p)$ and space $O(n^q)$, with $1 \le q \le p$.

Proof: 6.6.3: By Lemma 6.6.2 and Lemma 6.6.3.

6.7. Conclusions and Further Work

In this paper, we have introduced a version of safe recursion, together with constructive diagonalization; by means of these two operators, we've been able to define a hierarchy of classes of programs T_λ, with $0 \le \lambda < \omega^\omega$. Each finite level k of the hierarchy characterizes the register machines computing their output within time $O(n^k)$; using the natural definition of structured ordinals, and combining it with the diagonalization operator, we have that the transfinite levels of the hierarchy characterize the classes of register machine computing their output within time complexity bounded by the slow-growing function $B_\lambda(n)$, up to the machines with exponential time complexity. In the last section, we have defined a hierarchy of programs with simultaneous time and space complexity bound.

While the safe recursion scheme has been studied thoroughly, we feel that the diagonalization operator as presented in this work, or as in Marion's approach (see [12]), deserves a more accurate analysis. In particular, we believe that it can be considered as predicative as the safe recursion, and that it could be used to stretch the hierarchy of programs in order to capture the low Grzegorczyk classes (see [17] for a non-constructive approach).

References

[1]. E. Covino, G. Pani, A Specialized Recursive Language for Capturing Time-Space Complexity Classes, in *Proceedings of the 6th International Conference on Computational Logics, Algebras, Programming, Tools, and Benchmarking, (COMPUTATION TOOLS 2015)*, Nice, France, 2015, pp. 8–13.

[2]. A. Cobham, The intrinsic computational difficulty of functions, in Y. Bar-Hillel (Ed.), in *Proceedings of the International Conference on Logic, Methodology, and Philosophy of Science*, Amsterdam, North-Holland, 1962, pp. 24–30.

[3]. H. Simmons, The realm of primitive recursion, *Arch. Math. Logic*, Vol. 27, Issue 2, 1988, pp. 177–188.

[4]. S. Bellantoni, S. Cook, A New Recursion-Theoretic Characterization of The Polytime Functions, *Computational Complexity*, Vol. 2, Issue 2, 1992, pp. 97–110.

[5]. D. Leivant, A foundational delineation of computational feasibility, in *Proceedings of the 6th Annual IEEE Symposium on Logic in Computer Science, (LICS'91)*, Amsterdam, 1991, pp. 2–18.

[6]. D. Leivant, Predicative recurrence and computational complexity I: word recurrence and polytime, in Feasible Mathematics II, P. Clote and J. Remmel (Eds.), *Birkauser*, 1994, pp. 320–343.

[7]. D. Leivant, J.-Y. Marion, Ramified recurrence and computational complexity II: substitution and polyspace, in Computer Science Logic, J. Tiuryn and L. Pocholsky (Eds.), LNCS 933, *Springer Berlin Heidelberg*, Amsterdam, 1995, pp. 486–500.

[8]. I. Oitavem, New recursive characterization of the elementary functions and the functions computable in polynomial space, *Revista Matematica de la Univaersidad Complutense de Madrid*, Vol. 10, Issue 1, 1997, pp. 109-125.

[9]. P. Clote, A time-space hierarchy between polynomial time and polynomial space, *Mathematical Systems Theory*, Vol. 25, Issue 2, 1992, pp. 77–92.

[10]. D. Leivant, Stratified functional programs and computational complexity, in *Proceedings of the 20th Annual ACM SIGPLAN-SIGACT Symposium on Principles of Programming Languages, (POPL'93)*, Charleston, 1993, pp. 325–333.

[11]. M. Fairtlough, S. Weiner, Hierarchies of provably recursive functions, Chapter 3, in Handbook of Proof Theory, *Elsevier*, Amsterdam, Vol. 137, 1998, pp. 149–207.

[12]. J.-Y. Marion, On tiered small jump operators, *Logical Methods in Computer Science*, Vol. 5, No. 1, 2009.

[13]. T. Arai, N. Eguchi, A new function algebra of EXPTIME functions by safe nested recursion, *ACM Transactions on Computational Logic*, Vol. 10, Issue 4, 2009, pp. 24-1 – 24-19.

[14]. D. Leivant, Ramified recurrence and computational complexity III: higher type recurrence and elementary complexity, *Annals of Pure and Applied Logic*, Vol. 96, Issue 1-3, 1999, pp. 209–229.

[15]. S. Caporaso, G. Pani, E. Covino, A predicative approach to the classification problem, *Journal of Functional Programming*, Vol. 11, 2001, pp. 95–116.

[16]. M. Hofmann, Linear types and non-size-increasing polynomial time computation, in *Proceedings of the 14ᵗʰ Symposium on Logic in Computer Science, (LICS'99)*, Trento, Italy, 1999, pp. 464–473.

[17]. S. Bellantoni, K. Niggl, Ranking primitive recursion: the low Grzegorczyk classes revisited, *SIAM Journal on Computing*, Vol. 29, No. 2, 1999, pp. 401–415.

7.

Games as Actors: Interaction, Play, Design and Actor Network Theory

Jari Due Jessen and Carsten Jessen

7.1. Introduction

Using computer software usually implies that the human user is the active part who controls the interaction by input and direct manipulation [1-2]. Interaction with computer games is a different experience because the user acts in a game world where the contents of the game has extensive influence on the game player's behavior. Game figures and other game items are not just passive objects that can be manipulated by the player. For a game to come live, players must follow rules and act as the game requires. Playing a computer game like *Counter Strike* [3] or *World of Warcraft* [4] is not just a question of manipulating an avatar. The game is forcing the player to respond to events in the game by acting in a certain way if he wants to survive and prosper in the game, i.e. the player is placed in a role he must fulfill. In other words: games do something to and with people who play them and in that sense games are just like human actors who have an agency. What this agency consists of and how it is engineered is of interest to designers.

In this article, we will demonstrate how games can be regarded as actors and as organizers of actors and actions based on Actor Network Theory (abbreviated to "ANT") [5]. ANT is well suited for the analysis of user's interaction with games since ANT offers an approach to agency that does not assign power only to human actors but allows the possibility for objects and rules to be examined as actors. Also, ANT opens the door to viewing design as a social enterprise. As Yaneva stresses: "...design has a social goal and mobilizes social means to achieve it" [6].

ANT has received some attention in game studies during the last decade. Several scholars have studied games on the basis of ANT [7], focusing

Jari Due Jessen
Center for Playware, Technical University of   Denmark

primarily on the interchange between humans and technology [8] or on the development of social networks in online games and different physical environments take a different approach and show how the ANT perspective can explain which forces are at work while games are actually played. Our focus is thus on defining the immediate effects of using games. Our approach is in line with Seth Giddings [9], who have analysed games from the perspective of ANT. Giddings stresses that "the analysis of video games [...] demands the description of a special category of nonhumans, software entities ([...] agents) that act more or less autonomously or effect emergent behaviour" [9].

Our article is the result of a research project where we studied players of different ages playing computer games, board games, and digital play equipment. Contrary to Giddings and other scholars studying computer games, our point of departure was the theory that all games based on digital technology are games before they are anything else [10]. They are not first and foremost technology. Therefore, our study is focused on studying games as a genre rather than just digital games, and our main example here is a board game.

In the next section, we will introduce ANT focusing mainly on the concept of "translation" which is employed as our main analytical foundation. After this, the paper will present the research methodology applied for collecting data. In the following sections, the selected case of game playing will be presented followed by a presentation and a discussion of the results of our investigation. In this section, we will also draw on modern play theory to further explain how and why games functions and also why computer games belong to the general genre of games. We conclude this article with reflections on how our viewpoint may influence design.

7.2. Actor Network Theory

ANT was first developed by science and technology study scholars Michael Callon and Bruno Latour [11] as a new approach to social theory. ANT is of interest to any analysis of technology which goes beyond the assumption that technology is a mere instrument that we, as humans, utilize. ANT claims that any element of the material and social world (nature, technology, and social rules) can be an actor in the same way humans are. Agency is never only human or social but always a combination of human, social, and technological elements [12-14].

ANT is not a theory in the usual sense of the word according to Latour himself, since ANT does not explain "why" a network takes a certain form or "how" this happens [5]. ANT is more a method of how to explore and describe relations in a pragmatic manner, a "how-to book" as Latour defines it [5], and, as such, it offers a method to describe ties and forces within a network.

The main idea of ANT is that actions always take place in interaction between actors in networks where actors influence each other and struggle for power. We usually see social interaction between humans this way, however, ANT differs from traditional social theory by stating that actors can be any other elements as well.

7.2.1. The Traffic Example

ANT can be difficult to grasp and even counter-intuitive [12] because it reverses our common understanding of actors and agency, e.g. when it cuts across the subject-object division underlying our thinking about the world we live in. In an attempt to clarify ANT, Hanseth and Monteiro [15] use traffic as an example that explains the implications of seeing something in the perspective of ANT. We find their example very useful in obtaining a better understanding of ANT and, hopefully, what we later will write about what games do.

The following is a short presentation of their attempt and afterwards we will use it to explain the process of translation: When you are driving in your car from one place to another, you are acting, however, your acts are heavily influenced by technology, the material world (maneuvering abilities of the car, layout of roads, traffic signs, traffic regulation, etc.), and the immaterial (traffic rules, traffic culture, etc.) and habits (your own behaviour as a driver) [15].

According to ANT, these factors (including you) all function as actors and should be understood as forces of agency in a linked network where human and non-human, technical and non-technical elements are part of the network, and none of these elements are per definition granted special power over the others [12, 15].

Expanding the thoughts of Hanseth and Monteiro, we can add that, in the traffic example, you want to move from place to place, but you are dependent upon technology and forced to act in accordance with both social rules and physical conditions. Even though you are the driver, you will clearly feel the forces of other actors when acting out the driving. For instance, the road forces you to follow a certain route, the traffic light forces you to stop and start. One can say that in order to reach your goal safely and quickly, you have to "give in" to the network and in a way "hand over" your acting power and control over the car, so that the vehicle will move in accordance with the demands of traffic network. You must "delegate" [12] power to the traffic network, and, in return, you will reach your goal as fast and safely as possible. Of course, you are not actually handing over the control of yourself to the network. To delegate is more to act as prescribed and demanded by other actors. According to ANT, this is what happens in an actor-network relation and the purpose of an ANT analysis is to plot and describe the forces that are acting in the network of interest.

7.2.2. Translation

The way delegation is done is through the process of *translation*. This process requires the actors in a network to accept roles, a worldview, rules of acting, a path to follow etc. Michel Callon [16] describes the process of translation as a process of "persuading" with four distinct phases, he calls "moments": problematization, interessement, enrolment, and mobilization. These moments are inter-related overlapping steps that describe how stable actor-networks come to be established [17]. We will introduce them briefly in the following, and later use them in our game analysis.

The first moment, problematization, is where some of the actors in the network in question bring forth a definition of the problem and present a viable solution to it for the other actors. This is also the process during which the actors' roles are defined (both human and non-human actors). To use the example above, this is where the car, the traffic network, and traffic rules are presented as a solution to the transport problem, which is how to get safely from one place to another.

As part of the problematization process, a so-called obligatory passage point (OPP) is defined, i.e. a practicable solution, which the actors must accept to achieve their goal. An OPP "is viewed as the solution to a problem in terms of the resources available to the actant [actor] that proposes it as the OPP (...). It controls the resources needed to achieve the actant's outcome" [18]. By defining an OPP, other possibilities are closed [16]. In the traffic example, the OPP is literally a passage, since it's the roads and the current traffic rules, etc., which have been established as a solid, reliable network through a relatively long process that started when cars replaced horses as transport vehicles.

The second moment, interessement, is where the main objective is to persuade all the involved actors that the proposed problem and solution is the correct one so that they will accept to use this solution and not another one. In the traffic network this is done quite clear by authorizing traffic rules. In many networks intressement is a much subtler process, as we will see later in the analysis of games.

When the interessement of the actors is successful, the third moment, enrollment, is happening. This moment is important since it is here the actors become part of a network. The process can happen in many ways: "To describe enrollment is [...] to describe the multilateral negotiations, trials of strength and tricks that accompany the interessements and enable them to succeed" [16]. In relation to the traffic network, one can think of all the things that support cars and their moving in accordance with traffic rules along the roads and learning process human actors must go through to get a driver's license

Finally, the last moment, mobilization, is where the actors are mobilized in such a way that they act in accordance with their prescribed roles and thereby maintain the established network. This happens when the drivers actually drive their cars following the rules and pathways of the traffic network.

7.2.3. Design as Inscription

The effect of translation is delegation of power and agency. In relation to design of objects, e.g., computer games, translation differs from the traffic example with clear rules, since it is about how to construct an object in such a way that users are convinced to delegate agency voluntarily. Madeleine Akrich [19] has developed two useful concepts that describes translation as a process of *inscription* and *description*.

Inscription is the process where a designer embeds a special way the user must interact with the designed object. The designer is envisaging a user and a use case for the object and develops an intended use, which is inscribed into the object by use of, for instance, physical shape, GUI, behavior of objects, and affordances in general.

Akrich compares inscription with a movie script and calls the result a script for how the user should use the object. We see this in the design of e.g. the user interface of an iPad, where users are compelled to use finger movements to interact, which is a more intuitive way of interacting wand quite different from using a computer mouse.

While inscription is the designer's idea and framing of the interaction, Akrich uses the term description to describe the actual usage of the objects. This is where the script, built into and drawn upon in the design process, meets the user in an actual user setting. Coming alive is the central part of description. It is central to ANT that a non-human actor can have agency and perform actions and this is what we see when scripts, embedded in designed objects, come to life and objects engage in a network with other actors.

In the perspective of ANT, a game can be studied as a designed object with inscriptions that has agency and does something with the user, because the user invokes a network of actors and agency when he starts playing a game, i.e. following the rules of the "game world". A game designer should be aware of the network of actors that the specific game design can invoke if he wants to be able to use it in the process of inscription. Networks of actors represent the unit of analysis in our study presented below.

7.3. Research Methodology

Our research method relied on qualitative data collected through observation, based on non-participatory observation as well as active participation and interviews [20-21]. We collected data from 12 game sessions during which we observed informants, recorded their behavior and interviewed them before, during, and after playing. To ensure recordable data, we used games that required players had to be social and communicate with one another, and board games was particularly well suited for this since people tend to talk more when playing such games. We observed children as well as grown-ups and mixed age groups playing games in natural settings at home in a family situation or with friends. We recorded spoken language as well as body language and managed the many data using thematic and theoretical coding as described by Uwe Flick [22], who is inspired by Grounded Theory [23]. The analysis of the collected data was of course done using ANT. Researchers from social science have demonstrated that ANT is well suited for exploratory research in areas that have not been investigated much, not least because ANT-driven research is often able to draw up new conclusions [17, 24-25].

The purpose of our study was to investigate and describe agency and actors at work when players play games. As our framework of analysis, we employed the concept of actors and agency and the four described moments of translation, being careful not to differentiate between non-human and human actors. We analyzed agency by following what people did with games, extracting actors and ties, and described the translation process in the actual game situations, as we will demonstrate in the next two sections. These sections are also reports of findings from our study. As Kraal [17] writes with reference to one of the founding fathers of ANT alongside Latour, John Law: "It is the nature of ANT that it is easier to describe through a demonstration of its use".

It is important to mention that the object of our study is not the games themselves, but the *event* that unfolds when games are played [9]. In accordance with ANT, we analyse games in action when the forces of the network are at work, so to speak.

7.4. Case: The Game "Quackle"

The case of playing the board game "Quackle" in a mixed age group is used as an example for our observations in general and in the following, we will use our analysis of this case to present our interpretation of what the game actually does.

7.4.1. Quackle! Explained

The game, which was awarded "Game of the Year" in Denmark in 2006, is a typical funny board game for humans aged 5 and above. In short, the game consists of 12 different animal figures, 8 barns, and 97 playing cards with pictures of the animals and one arrow card (see Fig. 7.1). The game starts with each player pulling an animal figure from a cloth bag showing it to the others and then hiding it in his barn so the others can no longer see it. The cards are dealt and placed in a pile in front of each player face down.

Fig. 7.1. Photo of the game Quackle! In the box to the right are the barns (on top), the animals and the cards.

The objective of the game is to get rid of all the cards you have in your pile. Each round of the game consists of the players in turn turning a card and placing it for all to see. If two players have the same animal on their card they enter a *battle* during which the players compete on being the first to loudly say the sound of the *other* player's animal hidden in the barn. The player who loses the battle must pick his own and the pile of upwards facing cards of his opponent. The game continues until once again there are two identical animals in the cards or one of the players gets rid of all his cards [25].

The game seems pretty simple, but requires that the players can remember and quickly mobilize the correct sounds when two identical cards are turned, which is more difficult than one might think, even for adults.

7.4.2. Game Inscription

As we see in the above description of the game, there is a special way players are expected to interact with the game (the inscription) and, as we will argue in the following, in this way the game uses the learned scripts that the player brings along as well as physical and psychological abilities of the player. Among other things, the game takes advantage of the knowledge of the players (i.e. scripts) about animals and animal sounds, and the game utilizes the fact that most humans tend to react automatically in rushed situations. It is precisely this automatic reaction that makes the game funny, because the players make lots of mistakes trying to be the fastest, which often result in weird sounds that is a mix between different animal sounds.

The game designer has created an inscription that can be indicated as follows: We must say a particular animal sound while we see and try to remember a lot of other animals. These many inputs are combined with the stress factor that the game introduces by stating we must respond faster than our opponents! Thus, the inscription creates a special way the player must act, i.e. a specific way the players have to use their abilities.

In the perspective of agency, it is noteworthy that the game forces the player to make mistakes and thereby produce a mishmash of sounds, which he would not normally produce. When asking our informants about the experience, most of them said their tongue was "out of control". In this sense, it is evident that the game has agency and does something to the player.

7.4.3. Translation

The inscription plays an important role when considering the whole situation as a translation. As previously described, the translation consists of four moments which we will now outline in relation to the game scenario.

The first moment is the problematization, which is where we are presented with a problem. In our case, the game is played in natural situations on a Friday evening in a family of four (parents and two children, son aged 12 and daughter 21). For the family, the problem is the need for entertainment understood as a peaceful and enjoyable social time together. In this case, the game of Quackle is set up as a solution. Like any family game and most entertainment products, it promises that playing the game will lead to the experience of fun. Thus, the game is put forward as an actor who can do a piece of work (give us fun) through the way other actors treat it. This happens when one of the family members says, "Let's play Quackle, its fun. We always laugh so much when we play it" (quote from the daughter in this case).

The game is put forward as a solution and as the obligatory passage point (OPP) to social entertainment. The solution simultaneously suggests roles and organizes relations, i.e. a specific network where the family members will become game players and the living room table and chairs will facilitate the family sitting together. No less important is it that the game will establish equality between the players regardless of age and family position.

In the next moment, the interessement, which actually takes place in parallel with the problematization, the family members are convinced the proposed solution is the right one and barriers for alternative solutions to the problem are added. One of the things that are cut off is television, a frequently used source of entertainment in the family, when one of the adults says: "We shouldn't watch television, we always do. We should do something together instead" (quote from the episode).

Enrollment is the third moment where the players are enrolled and this entails that they must accept the roles of participants as players of Quackle and accept the terms of the game. Since the family have played the game of "Quakle!" before, they do not need to learn the rules, which in other cases often is part of the process of enrollment.

In the last part of translation, mobilization, the solution is executed when the family members sit down with the game and start playing. If the mobilization works and the translation process is thus successful, it enables the family to experience fun and laugh together. This is exactly what happened to the test family via the interaction with the game, which created a lot of laughing especially when the parents made weird sounds.

In our observations, we also encountered an event of a failed translation. In this episode, which involved four adults and two children, the setup was similar to the above but the one of the players did not accept the role of a player who could end up saying a weird, funny sound, and thus she ended up destroying the game. She did not hand over agency to the game and did not act as prescribed by the game. This episode was special, but its points to the fact that the translation can fail and the participants have a choice, though this choice comes with certain consequences (they never got in to play.

Going back to the situation with the successful translation, the game re-organizes the social connections within the family and in so doing builds a new network of actors and agency. The game is what Latour has named a "mediator" that "transforms, translates, distorts, and modifies" relations [12]. But the game does more than alter the social relations. It mediates the body and mind of the individual players. In the following, we will address how Quackle accomplishes the mobilization of the physical and cognitive abilities of the play.

7.4.4. What the Game Does

A game cannot do much by itself but is dependent on other actors, and this is, of course, particularly true for board games. Nevertheless, games have agency that makes game players act in a manner they would not have acted without the game. In that sense, the game "does" something in line with Latour's concise statement on what defines an actor: "anything that does modify a state of affairs by making a difference is an actor [...]" [5].

Latour stresses that when we are studying a network in ANT, we are focusing on the circulation between the connections that make up the network [17]. When we study the Quackle game, we are looking at how agency is floating between the involved actors, the details of which we will try to demonstrate through an analysis of a play scenario. First, the scenario of a family playing the game:

1. The game is placed on the table and the players sit down around it.

2. The game is opened, and the game elements are displayed. There are animals, barns, and cards and a cloth bag.

3. The animals are hidden in a cloth bag and all players get a barn.

4. Each player pulls an animal from the cloth bag: Player 1 gets a snake, player 2 a dog, player 3 a donkey and player 4 a frog.

5. After all animals and sounds have been introduced, they are stored out of view in the barns.

6. The cards are shuffled and dealt.

7. Everyone is ready and turn their first card.

8. A horse, a cow, a duck and a pig is turned, so there is no match.

9. Next cards are turned: a snake, a pig, a frog and an owl appears, still no match.

10. The third cards are turned: A mouse, a donkey, a rooster and an owl appear.

11. The game gathers speed and the cards are turned a bit faster.

12. The fourth card is turned: a cat, a dog, a cat and a frog.

13. Player 1 shouts "Qu..iau" [sounds a combination of a frog sound and a cat sound] and player 3 "Vuu..shh"[a combination of dog sound and snake sound] followed by a grinning "Oh no, uh" and finally player 1 says "Miau" just before player 3 says "Sssshh".

14. Player 3 must gather player 1's card and the game continues.

This is the basic structure of the game which continues in a similar manner for a long time (about 30 minutes) before one player wins. Points 1 and 2 are of practical character, but they help to create the framework for what is going to happen. Thus, the following activities are framed and the game's inscription starts to become clear, especially in the form of the rules. The agency is still with the players. This is also the case in point 3, but here the game starts to gain agency. It starts to influence the players, as it prescribes their actions in the next steps.

Our observations show that at the same time the players build up anticipation about what is going to happen, which is seen by the body movements and heard by the tone and pitch of voices. This anticipation started when the players accepted the game as an OPP, that is when they all agreed upon plating the game. It was especially noticeable in points 4 and 5, where the joy of hiding the animals in the cloth bag and pulling one provides a form of excitement that is particularly evident in the youngest child. Thus, we see here that the agency is distributed to the game as a kind of pre-disposition of body and mind 6.

In point 5, the players need to remember all the animals the other players have. The individual player must establish links between the different animals and the other players around the table. In point 7, the number of links is expanded by the creation of connection to the cards and in point 9, the game is made even more complex as more animals are introduced and it makes it harder to remember the animals hidden in the barns, which is of course part of the game designer's inscription.

We continue to point 13, where we see the first match of cards. When this match appears, a special script starts taken over, which is part of the inscription in the game. The script forces the player to act as prescribed by the game rules and it thereby functions as a type of mechanism that governs the actions of the players. What especially interesting here is that this mechanism *re-organizes* the normal connection between the player's body, mind, and cognition in a special way by means of rules and materials (cards, animal figures, barns) and, in this manner, the game utilizes some of the players abilities. As mentioned earlier, the player is driven to make mistakes when pronouncing words, and it is this "drive" that demonstrates an agency from the game.

What the game does can be described as follows: First, it mobilizes the individual player's memory but overemphasizes the need to remember. There is a wide range of images, sounds, figures, and places in play, and the player will have to revive all of these objects and connections when the match of cards happens. There are different animal figures and their sounds to choose from, and several sounds usually become actualized before the players can produce the correct sound.

Secondly, the game cuts across the usual connection between the player's mind and body. In point 13, it is clear that the game disrupts the usually well-controlled connections between the players cognitive ability and their ability to control their voice and words. The inscription provides a procedure for a specific requested response to certain signals where the player must use specific cognitive functions, i.e. perceive, remember, associate images and sounds as well as mobilize the organs of speech; and it all has to happen as quickly as possible. It is a simple task that players do not usually have problems with, but by adding a wide range of signals in the form of different images and sounds, and by forcing the players to compete with each other, the result is that cognitive and bodily functions respond in an incorrect manner and the players end up making wrong sounds. The game has, in a way, taken over body and mind.

The case of playing Quackle is an example of a translation process in action, where agency is delegated to a network. The case is also an example of how such a network is comprised of human, material, and social actors. The translation is only happening because the players have allowed themselves to be enrolled as players and fulfill their roles by using the material and following the rules and thereby delegating agency. In return, they are entertained.

7.5. Playing a Computer Game

Earlier in this article, we stated that we consider computer games to be games before anything else. Thus, our thesis is that computer games do something to the players when played, just as in the case of Quackle. What we have attempted until now is to establish a framework for analyzing what games do, and, in the following, we will show how the framework can be applied to computer games.

The setting, which we observed, are three boys 12, 12 and 14 years old playing Grand Theft Auto V (GTA) on a Playstation. GTA has become very popular with its mixture of racing and adventure, where the players can follow a story already inscribed in the game or they can just go racing around in the game city (Fig. 7.2).

Fig. 7.2. An example of driver's view i GTA.

The boys take turns at controlling the game with the game controller while the two others comment and talk about what is happening. In one scenario, the 14 years old has the game controller in his hands, when he gets an assignment from the game. A tough looking guy on the screen tells him that he needs to win a race with a computer-controlled opponent to progress. Then the game begins.

The setting, we are analyzing, is a network that consists of the interior (couch, table, etc.), the Playstation (consisting of screen, game console, controller and DVD), the three boys, and the game. The game itself consists of multiple actors of which some are activated in combination with the other actors of the network.

We will not analyze all actors and possible networks the game can initiate but will only take a look at how the game impacts the players' bodies.

When playing, the boys must of course follow the rules of the game. They are complicated, but for our example here we can just point to the traffic rules in the game and how the car is driven via the game controller. In the same manner as in a real traffic system, the player has to delegate agency to the system. Just as in the real traffic, there is police, here in the form of multiple cars and helicopters, and there are roads, houses, pedestrians, and the normal traffic on the road, all of which have to be avoided during the race. These actors become active as the boy starts the race, which lasts for a few minutes.

It is apparent how the game influences the player's body. To initiate the game, the boy presses hard on the controller and swings it forward, and the next second he and the controller are leaning heavily to the left side, almost leaning into one of the other boys. The next second, all of the boys shout "Wow, that was close!", while they all jump a little in the couch. At the end, they are all standing up and leaning forward and to the side as they follow the movements of the car on the road it tries to follow.

If we look at this scenario as a translation, we can see the problematization is set forward as the boys need to win the race and this is also the OPP. In the interessement, the game builds on the fact that the boys are already enrolled in the game (emerged in it) and thus they need to progress to keep playing. The enrollment is made more stable by the use of a character in the game (an avatar) and adding a storyline to the race, which tells why the player has to win. Thus, agency is transferred to the game. This also builds up the tension for the next moment, where the boys are mobilized to play. The term "boys" indicates that all three boys participated with movements and speech even though two of them did not control the game.

When the race begins, the boy controlling the game is leaning forward and swinging to the side with his body. This is where the game uses some of its agency and the bodily action of the player shows that the game is mobilizing some of the player's abilities, in this case his knowledge of spatial movements, which he unconsciously carries with him. In our observations, we saw this again and again, the players could not help it but move their body to the side as the car in the game turned a corner, even though in this game it was not needed, as the controller does not react to it.

The game use mental pressure in much the same way as do Quackle to take agency when it forces the player to react to a lot of events that constantly appears when the player's car on the screen meets objects in the game, like other cars or pedestrians that must be avoided.

The game further use agency over the human actors when it makes the boys shout and jump. This happens as the player's car almost hits a wall that would have crushed his car and made him lose the game. This kind of danger is present all the time in the race. Here, the game is exercising its agency by using the player's body and mind, including his imagination that allows him and the other boys to experience a danger, which in the real world would have produced fear but, in the framework of the game, produces excitement and joy.

7.6. Theory of Play and Games

Obviously, excitement or pleasure is the reason why game players obey to the demands of games in the way we have described above, i.e., accept to act as a node in a network, following rules they sometimes do not understand, and often using hour after hour trying to learn to manage computer game challenges. What games do is to produce play and playful experiences by utilizing the players body and mind. In the following, we will lean on modern play theory and modern game studies to clarify the importance of play and the connection between games and play.

One need not search for long in game studies literature before it becomes evident that play, according to most researchers, is an important factor for the success of computer games as well as other kind of games. Prominent play scholars like Johan Huizinga, Roger Callois, Gadamer, and Brian Sutton-Smith appear as references in numerous articles and books on the topic. In Salen and Zimmerman's well know book on games, *Rules of Play* [27], the authors define the goal of successful game design as "...the creation of meaningful play..." [27] and later on state that "...rules are merely the means for creating play..." [27]. And to make the central point absolutely clear, they argue in a subsequent anthology on games that "...games create play: of that there is no doubt" [28]. In other words, games fulfill a function in relation to play.

In line with our view presented here is also [29-31]. Games can be regarded as "tools" that generate play, and, more importantly, games must be designed with the aim of creating play among the users.

But what is play? In developmental psychology, play is primarily seen as a means for learning (Piaget [32], Vygotsky [33], Singer, Golinkoff, & Hirsh-Pasek [34]) and, in that frame of reference, it follows logically from the statement that games generate play that they also generate learning. Modern play theory understands play differently. Based on the work of the above-mentioned play scholars, play is regarded, in and by itself, as a meaningful human activity that we practice for the simple joy of it. In that perspective learning can still take place, nonetheless learning is not the reason or motive for engaging in play activities, but a byproduct. Game players accept the translation of agency to games simply because they can get into play by doing so, or more accurate get into the condition in play theory called "the state of play", derived from Johan Huizinga [35] who is probably the most quoted play theoretician today. He writes in "Homo Ludens" (which translates to "man, the player") about play this way:

"...what actually is the fun of playing? Why does the baby crow with pleasure? Why does the gambler lose himself in his passion? Why is a huge crowd roused to frenzy by a football match? This intensity of, and absorption in, play finds no explanation in biological analysis. And yet in this intensity, this absorption, this power of maddening, lies the very essence, the primordial quality of play. [...] ... it is precisely the fun-element that characterizes the essence of play. Here we have to do with an absolutely primary category of life, familiar to everybody. [...] the fun of playing resists all analysis, all logical interpretation..." [35].

The last sentence is perhaps the most important for the understanding of play and, thus, for the understanding of what games should be designed for. Play is a difficult concept to define in a scientific context because of its nature as an activity, which represents other values than the ones we traditionally use and base our thoughts on. Both in science and in our daily lives, we usually try to rationalize human activities and give them a purpose. When it comes to play, it is not possible to apply rational reasoning according to Huizinga, and play does not submit itself to the usual rational notions. We are forced to remove our accustomed patterns of thoughts and recognize that the human being is something else and more than a rational being. In short: Human beings want to play for the fun of it, and we use games primarily because they can get us "absorbed" in play.

Games, whether board games, computer games or other kind of games (of which we will present an example shortly), should be designed to facilitate this absorption. Traditional games like street games that have been around for a long time, some for hundreds of years, are clearly designed to produce the joy of play [31]. Games are some of the first things we meet as infants when we learn to communicate. Play researcher Brian Sutton-Smith have given one of the most precise definitions of play, which is useful to game design, even if it is about infants:

"[…] we postulate as the aboriginal paradigm for play, mother and infant conjoined in an expressive communicational frame within which they contrastively participate in the modulation of excitement. We call this a paradigm for all ludic action, because we suggest that other play itself is a metaphoric statement of this literal state of affairs. Ludic action, wherever it is, always involves the analogous establishment of the secure communicational frame and the manipulation of excitement arousal through contrastive actions within that frame" [36].

"Modulation of excitement" is a very accurate description of what games do. There are numerous variations of such modulation. For instance, play can be physical, making the body move forward and backwards, as in sports, dancing, or on a swing; it might be psychological, creating and using a mental tension, for which jokes or horror stories are good examples. It is remarkable in this context that play is often generated by directly using the natural reactions of the body and mind, e.g. confusion due to overemphasizing of information in Quackle or fear in GTA, as we have tried to show in our game analysis. Perhaps the most simple and recognizable example is twirling that produces an enjoyable feeling of dizziness. For small children, it is enough to turn around again and again to get that feeling while grownups must attend a fairground and pay for muse rides to achieve the same. Still, the dizziness is similar, and a similar pointless waste of time, if it was not for the mere fun of it.

We employ countless forms of materials, techniques, or genres of physical as well as immaterial types to help initiate activities that make us play. Thus, games are just one out of numerous tools ([27, 29-31]). From the simplest tools like twirling, to the computer games the goal contains a familiarity. In the next section, we will present physical games built on high tech equipment, where we have utilized knowledge of games as tools for play.

7.7. Exergames

Exergames is one of the many names for a new type of games. These games try to combine physical exercise with digital games through an interface that requires physical exertion [38-39].

Exergames are interesting here because they combine the physical abilities of the players with the special opportunities of digital games. At the same time, many of these games are less complicated than computer games like GTA, because they rely on the physical aspects and movements of human players and less on the virtual world's narratives. This allows us to further investigate how the human players abilities are being used within the network of a game.

In the following, we will consider one type of exergaming called Moto Tiles consisting of modular interactive tiles ("tiles" for short).

The tiles (displayed in Fig. 7.3) are a distributed system consisting of electronic tiles, which can be assembled like puzzle pieces. The tiles combine robotics, modern artificial intelligence, and play in a product that can be used for games, sports, health promotion, rehabilitation, dance, art, and learning [39].

Every tile is 30 × 30 cm and works independently but can communicate with all the surrounding tiles. In this way, all the tiles can communicate with each other and create a playfield for the players to play on. The tiles have a force-sensitive resistor and eight RGB light-emitting diodes able to shine in a rainbow of colors.

The many colors allow for a variety of different types of patterns and games to be played. To play a game on the tile platform, a player must move around and step on the tiles according to the rules of each game (see later). The various applications can either be played by a single person or can be set up so that multiple people can collaborate or compete against each other.

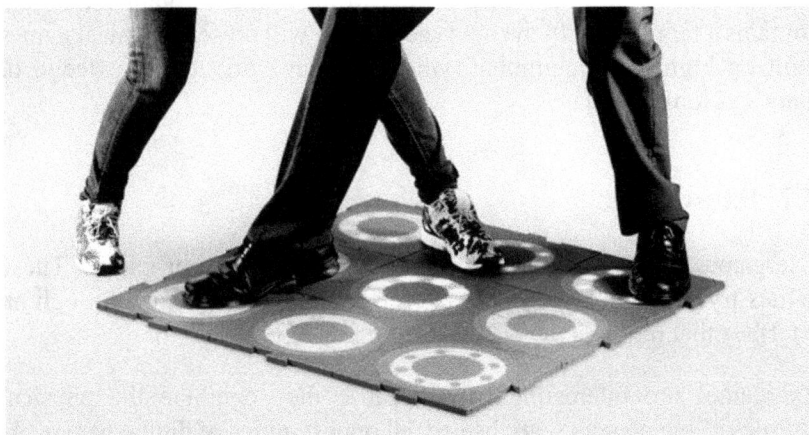

Fig. 7.3. Modular interactive tiles invented and produces by Center for Playware, Technical University of Denmark.

Because the tiles are designed to work in any combination and because of the puzzle piece design, the tiles give the user the ability to create any playing field they wish, and to change it again anytime - e.g., change the size or shape of the field of tiles. When the user changes the configuration of the tiles, the interaction and difficulty is also changed, e.g., faster/slower movements, longer/shorter steps and so on. Thus, the user can change the movement and difficulty merely by physically building a different kind or size of the platform.

The tiles have been used as balance training for elderly people (65+ years old) and motor skills training for children (5-6 years old). We observed both elderly and children (total of 20 sessions for each group), but here we will focus on the sessions for children. Each participant partook in 10 or more sessions and a total of 19 children participated.

The data were analyzed using the methods described earlier and the following account is a prototypical example of the use of the tiles for children even though similarities exist in the use for the elderly. This example illustrates the main findings and forms a good basis for the ANT analysis. For the sake of the analysis, we are focusing on one game called "Color Race", (see [39] for more info on the Moto tiles).

The game "Color Race" is a type of "Catch the Color" game. On the playing field, three tiles are randomly displaying different colors – red, green, and blue. Each player chose a color and must step on the tile with the chosen color as fast possible. When they step on the tile, its color shifts randomly to another tile on the playing field that the player now must step on.

The player must step on tiles with their chosen color as many times as possible within a given timeframe (typically 30-60 sec). When the time is up, all tiles light up in the color that got most points. Hopefully, the reader can imagine three players running around on the relatively small playing field at the same time trying to step on tiles as fast as possible. The stage is set for rough-and-tumble play (in our experience regardless of age).

In the scenario that we believe is a prototypical example of the use of the tiles, we are in a kindergarten with 10 children 5-6 years old and an adult. The room is full of other toys, but there is room in the middle of the floor for the tiles. They also have chairs that some of the children are sitting on while they are waiting to play. Others are standing around and cheering or observing. The children are playing on the tiles two times a week, so they know them at this point. The adult helps to set up the tiles and they are placed in a typical setup of 9 tiles in a 3×3 square, and the game Color Race with three colors is started. Three of the children pick a color each, and they place themselves in front of that color and countdown to start.

As soon as they start the game, they jump from tile to tile trying to get around the other players, but they keep bumping into each other again and again as the playing field is approx. 1x1 meter so they do not have much space to move on. The game lasts for 30 seconds where the players jump around and get around 20 points each. At the end of the 30 seconds, the tiles light up in green showing that the player, who picked green, got the most points.

As described above, the game requires the player to step on the tiles for the game to proceed. Here the inscription is the tiles in general that are calling for the player to step on them. In our observations, we have seen time after time that new players or observers, who have no knowledge of the tile in advance, can not resist trying to press the tiles to see what happens. The physical design of the tiles on the floor, the size of a foot, and the colorful light invite the players. They function as trigger points.

If we look at the inscription, it can be described as follows: The player must press a tile and catch as many lights as possible within a limited time frame. The game is created so the color jumps to another tile almost instantly and this creates the feeling of running after the colors, thus the name "Color Race". The movement of the light to another tile "forces" the player to act as prescribed be the game rules but also the surrounding network of competing with other children, and the observers cheering on is contributing to this "force". This is another example of what we saw earlier with Quackle!, where the players are driven to act in a certain way. Since each player must press a tile and catch as many lights as possible within a limited time frame, which organize both body and mind of the players and the interaction between them. The game also creates the necessity of speed by organizing the game as a competition. All the sessions we observed with children involved multiple players on the platform, and with more players at the same time, there is also an element of competition and a lot of communication, both linguistic and bodily, between players. It is noteworthy that all the players, we have observed, talk, shout and laugh. The game evokes a kind of friendly play fight.

It is of special interest from our viewpoint that the game sets up the individual players not only as players, but at the same time as material obstacles in the game. In the scenario with the tiles, the players are all playing at the same time and the colors jump around the platform. Here the game is using its agency. As pointed out above, the game is forcing the players to move from one tile to another, but in the process it creates a "double" role for them as players also become obstacles for other players. This "double" use of the human player is important for how the game functions. Each player becomes a game element, as they again and again are standing in the way of others who are trying to reach a tile with their color.

In the observations, we could see that exactly this point was critical for how much fun the participants seemed to get out of the game. If they surrendered to the game and accepted and maybe even used the fact that they bumped into each other, they seemed to enjoy the game more. Often players tried to push, pull or bump the other players away so they could easier reach a tile.

The game is also pushing the players to speed up and jump around by shifting the position of the light almost instantly as the tile with a color is pressed. It creates the effect of the game progressing fast, and players indicated that they felt the need to hurry to the next tile even though the color will stay there until pressed. Technically, there was no need to hurry but mentally it appeared so.

If we consider the case as a translation we can then observe the problematization as the case of the children wanting to get into play (the state of play), and the tiles are put up as the OPP. In the interessement, the children are convinced that the tiles are the solution to the problem and the roles are divided with the children as players and obstacles for each other, the tiles as the playground and the place the game will take place.

The children are enrolled when they accept their role the rules of the game, thus accepting that they will become both active players and obstacles in the game.

In the final moment the actual game is played. The children run around on the playing field and the tiles make them move from one tile to another, shifting their balance, running into each other, and fighting to get the most points and by doing that clearly producing the state of play.

In the case above, we tried to make it clear that the players can take multiple roles in the game, and that the actors of the network can be used both with their mental abilities (e.g. competitive revivals) as well as their physical or virtual manifestation (e.g., obstacles or trigger points).

In the following, we will go deeper into what the implications of these analyses of games in the view of ANT have for designers of games.

7.8. Design Implications

In the introduction, we stated that games, in our point of view, could be regarded as actors because they function as organizers of other actors. Following Latour, quoted above, games are actors because they make a difference; not because they are human or non-human, social or material. We have tried to show how such "difference" is created when games do something with players. This view represents an understanding of interaction where the subject-object dichotomy is dissolved and agency is distributed in a process of reorganization, recreation and modification of actions in networks that even stretch into the mind and body of the individual player and take advantage of human abilities and faculties.

If one accepts this way of viewing, it will have implications for game design, because design is not just a question of creating game worlds and interfaces but also a question of how to design social actors that can take agency and thereby initiate and guide the building of social networks, which can bring human and non-human actors to act together in such a way that the players can achieve an experience they find pleasant, joyful, funny or equivalent. As we have tried to point out, this does not only involve organizing social relations, actions and material, but also requires utilization of the player's abilities of both physical and cognitive nature.

We believe game design should be done based on knowledge about how human abilities can be organized and influenced including knowledge of the abilities of different user groups. In the analysis, we showed how games orchestrate actions by humans and non-humans and resulted in experiences the players find engaging, joyful, and entertaining. From our point of view that is prototypical examples of what games do. They organize the acting of actors in order to achieve certain kinds of experiences, which, as we have argued, primarily are states of play.

Through the inscription, the designer assigns agency in such a way that the game can take advantage of the characteristics of the human players. The games are examples of how the designer renders agency to a non-human object, and how these objects perform a job by getting the players to do a job. This view gives us a possibility to further investigate how the designer can utilize this understanding when creating games.

Understanding games or more precisely game elements as active participants in the network created by or around the game, puts emphasis on attributing agency to the game and the elements in it. To understand how this is done, the concept of framing from communication theory is useful.

Framing is a concept developed from observations of play by Gregory Bateson [40], who points to the fact that certain situations are perceived differently than we normally would in his essay with the title "This is play" [41] which is now famous both in the context of communication and play research.

The classical example from Bateson is two monkeys playing; where in this framing a bite (an act of attacking)) does not denote what it normally would (fighting against each other, trying to do harm) but is framed in such a way that it is perceived differently. Bateson states that a bite in the frame of play must be followed by a metacommunicative signal "this is play", so that the opponent understands it as an act in play and not seriously meant [40-41]. This is, for instance, the case with computer games such as GTA and the play fighting on Moto tiles that we have described earlier. "This is play" puts a frame around every act which signals "not serious". But that does not mean that the acts are without influence on the players. For our viewpoint, this is a tricky point which we must elaborate on.

The best example is perhaps the feeling of fear. Psychologist and play researcher Michael Apter [42] have put forward the example of meeting a tiger. There is a significant difference between meeting a tiger face to face in the backyard and meeting tiger in a cage, he writes in an attempt to explain that the way we experience our surroundings changes their significance due to the frame we put them into. This is especially true in play. Events which outside of play would produce fear, anger and the like, does not produce the same reactions in the framework of play. Still, as the Apter example shows, what we experience in play must *evoke* some of the same feelings as in reality. If not, we would be bored. A kitten in a cage is not exciting but pitiful. We believe this is a key point in designing games. The "modulation of excitement" of course requires something to modulate. Fear is only one example. Apter writes:

"One of the most interesting things about play is the tremendous variety of devices, stratagems and techniques, which people can use to obtain the pleasure of, especially to achieve high arousal [...]. Putting aside the use of direct physiological interventions to increase arousal – drugs smoking and drinking – there are a number of general psychological strategies which can be used for this purpose" [42].

A designer must know which emotions, feelings, etc. that produce arousal or other kinds of excitement and joy in the specific target group, and must know how to evoke them in a game. Good designers know that by intuition; however, explicit knowledge may help to make games better or to better avoid failures.

In terms of a game taking agency, the key point is to set the scene for the game by creating a framing where the players are willing to invest time and energy in the game and in the process delegating agency to the game. The players also must accept the roles and rules of the game. Often this framing is done in the terms of narratives where the designer includes a story that frames the game and divides the roles. Dividing the roles and hereby building the social network is an important part of the work done by games. This is also the first part of the translation.

We described this in the case of Quackle and how it divided special roles. This is also especially clear in GTA and the case of the tiles. In GTA, the social network is built to include the actors of the race but also draw on the bigger picture of why players have to advance through the race. In the case of the tiles, the social network is constructed to create a social awareness of the actors and how they compete and play around with each other.

7.9. A Word on Scripts

The social networks and relations, actions, and materials are not the only elements to take into considerations. The most vital part that the ANT analysis points to, is to take the abilities, feelings and emotions of the players (physical as well as psychological) into account. As described earlier in the inscription, the designer can take advantage of the scripts that the players already have "downloaded", e.g. the fear of tigers, to mention a simple script.

In the example of Quackle, it was the ability to make the sounds of the animals combined with a common script that made us laugh when we and other people made mistakes inside the frame of play. In the case of the tiles, it was the game structure of "Color Race" where the players had to "catch the color" combined with the script of playful fight. Players know this kind of game; they know how it is played and the designer can use this knowledge.

All these examples are scripts of different types. As described earlier, scripts are a form of manuscripts that we know and which we use to interact and cope with different situations. In a sense, scripts can be seen as a form of recipes.

In that sense, games are dependent on the players. Players have many different scripts and understandings of how to play and what a game is. All these can be seen as part of their play culture. When players play a game or observe others playing, they learn new ways of playing and interacting: new scripts are passed to them.

It is sometimes easy to see, as when a child looks at elder children playing and starts to mimic their behavior. In this situation, the child is starting to "download" the script for their actions and can later reuse these.

In all these small scripts, we have learned that the designers of game are using them in different ways while they are at the same time supplying new ones to the players.

7.10. Conclusion and Future Work

The main theme of this paper has been to establish an understanding of what games do in the perspective of ANT. We have seen how games do an active job and work as what Latour calls a mediator that can "transform, translate, distort, and modify" relations [12]. We believe that ANT is beneficial when we look into computer game design. While it can seem trivial that games do something to users, it is highly important for game designers to understand how games do this and why people are willing to invest time and effort in games.

We have demonstrated that, using ANT as a tool for analysis, can give us a new understanding of the interaction between games and users. We believe that game designers can advance interaction design by "following the actors" and by understanding how agency in games works, and by gaining insight into how certain bodily, psychological, and social acts can create play. We are fully aware that our analysis has shortcomings since it only covers three games although several instances of them and, thus, only a few examples of the kind of actor network which creates play. There are numerous other examples of this kind of network operating in many different ways in games. Jessen and Lund [43] have developed the concept of "play dynamics" in an attempt to establish a new vocabulary for describing the countless phenomenon, which can create the state of play in humans. They define the concept as follows: "A play-dynamic is the dynamic effect of the play-force which affects the player by placing this person in a state of playing." The concept is in line with our understanding of how actor networks and scripts functions in games, and the concept "play forces" is a feasible way of describing how games utilizes the players mind and body. Future work should focus on identifying, characterising, and possibly systemizing actor networks in different games. It should focus on identifying different kinds of key scripts or play dynamics that designers can take advantage of when designing and evaluating games. Similarly, it's interesting to further investigate how the understanding of games as translation can help games theory to create a better awareness of what is going on when games are in use.

References

[1]. J. D. Jessen, C. Jessen, What games do, in *Proceedings of the 7th International Conference on Advances in Computer-Human Interactions (ACHI 2014)*, 2014, pp. 222-224.

[2]. P. Dourish, Where the Action is – The foundation of Embodied Interaction, *The MIT Press*, Cambridge, 2004.

[3]. Valve Corporation, Counter Strike, *Valve Corporation*, Washington, 2011.

[4]. Blizzard Entertainment, World of Warcraft, *Blizzard Entertainment*, Irvine, California, 2004.

[5]. B. Latour, Reassembling the social: an introduction to Actor–network theory, *University Press*, Oxford, 2005.

[6]. A. Yaneva, Making the Social Hold: Towards an Actor-Network Theory of Design, in *Design and Culture*, Vol. 1, No. 3, 2009, pp. 273-288.

[7]. M. Cypher, I. Richardson, An actor-network approach to games and virtual environments, in *Proceedings of the International Conference on Game Research and Development (CyberGames '06)*, 2006, pp. 254-259.

[8]. K. Kallio, F. Mäyrä, K. Kaipainen, At Least Nine Ways to Play: Approaching Gamer Mentalities, *Games and Culture*, Vol. 6, No. 4, 2011, pp. 327-353.

[9]. S. Giddings, Events and Collusions A Glossary for the Microethnography of Video Game Play, *Games and Culture*, Vol. 4, No. 2, 2009, pp 144-157.

[10]. U. Plesner, Researching Virtual Worlds: Methodologies for Studying Emergent Practices, *Routledge Studies in New Media and Cyberculture*, Vol. 14, 2013.

[11]. M. Callon, B. Latour, Unscrewing the Big Leviathan: how actors macrostructure reality and how sociologists help them to do so, in Advances in Social Theory and Methodology: Toward an Integration of Micro- and Macro-Sociologies, K. D. Knorr-Cetina and A. V. Cicourel (Eds.), *Routledge and Kegan Paul, Boston*, 1981, pp. 277-303.

[12]. B. Latour, The Trouble with Actor-Network Theory, in Om aktor-netværksteori. Nogle få afklaringer og mere end nogle få forviklinger, F. Olsen, Philosophia, Vol. 25, No. 3-4, 1996, pp. 47-64.

[13]. B. Latour, Where are the Missing Masses? The Sociology of a Few Mundane Artifacts, in Shaping Technology/Building Society, W. E. Bijker and J. Law (Eds.), *The MIT Presse*, Cambridge, 1992, pp. 225-259.

[14]. B. Latour, A Door Must be Either Open or Shut: A Little Philosophy of Techniques, in Technology and The Politics of Knowledge, A. Feenberg and A. Hannay (Eds.), *Indiana University Press*, Bloomington, 1995, pp. 272-281.

[15]. O Hanseth, E. Monteiro, Understanding Information Infrastructure, University of Oslo [online]. Available from: http://heim.ifi.uio.no/oleha/Publications/bok.html, 2014.01.16

[16]. M. Callon, Some Elements of a Sociology of Translation: Domestication of the Scallops and the Fishermen of St Brieuc Bay, in Power, Action and Belief: A New Sociology of Knowledge, J. Law (Eds.), *Routledge & Kegan Paul*, London, 1986.

[17]. B. J. Kraal, Actor-network inspired design research: Methodology and reflections, in *Proceedings of the International Association of Societies for Design Research Conference*, Hong Kong, 2007, pp. 1-12.

[18]. J. Rhodes, Using Actor-Network Theory to Trace an ICT (Telecenter) Implementation Trajectory, *Information Technologies & International Development*, Vol. 5, Issue 3, 2009, pp. 1-20.

[19]. M. Akrich, The Description of Technical Objects, in Shaping Technology/Building Society: Studies in Sociotechnical Change, W. Bijker and J. Law (Eds.), *The MIT Presse*, Cambridge, 1992, pp. 205-224.

[20]. J. P. Spradley, Participant Observation, Orlando, *Harcourt College Publishers*, Florida, 1980, pp. 58-62.

[21]. P. Atkinson, M. Hammersley, Ethnography and Participant Observation, in Handbook of Qualitative Research, N. K. Denzin and Y. S. Lincoln (Eds.), *SAGE Publications*, Thousand Oaks, 1994, pp. 248-261.

[22]. U. Flick, An Introduction to Qualitative Research, 3rd edition, London, *SAGE Publications,* Thousand Oaks, 2006.

[23]. Juliet Corbin, Anselm L. Strauss, Basics of Qualitative Research: Grounded Theory Procedures and Techniques, 3rd edition, *SAGE Publications*, 2008.

[24]. R. Dankert, Using Actor-Network Theory (ANT) doing research, in Publicaties Vanaf, 2010 [online]. Available at http://ritskedankert.nl/publicaties/2010/-item/using-actor-network-theory-ant-doing-research, 2014.01.12

[25]. R. Nimmo, Actor-network theory and methodology: social research in a more-than-humanworld, in *Methodological Innovations Online*, Vol. 6, No. 3, 2011, pp. 108-119.

[26]. Algaspel, Quacklemanual, *Algaspel*, Stockholm, 2011.

[27]. K. Salen, E. Zimmerman, Rules of play: game design fundamentals, *The MIT Presse*, Cambridge, 2004.

[28]. K. Salen, E. Zimmerman, The Game Design Reader: A Rules of Play Anthology, *The MIT Press*, Cambridge, 2005.

[29]. C. Jessen, Interpretive communities. The reception of computer games by children and the young, Odense University, 1999 [online]. Available at: http://www.carsten-jessen.dk/intercom.html, 2014.01.12.

[30]. H. S. Karoff, C. Jessen, New Play Culture and Playware, in *Proceedings for BIN 2008*, Copenhagen, 2008 [online]. Available at: http://vbn.aau.dk/files/73392625/-3BINjessenkaroff.pdf, 2014.01.15.

[31]. H. H. Lund, C. Jessen, Playware - intelligent technology for children's play. Technical report, Maersk Institute, University of Southern Denmark, 2005 [online]. Available here: http://www.carsten-jessen.dk/playware-article1.pdf, 2014.01.14.

[32]. J. Piaget, The psychology of the child, *Basic Books*, New York, 1972.

[33]. L. S. Vygotsky, Play and Its Role in the Mental Development of the Child, *Soviet Psychology*, Vol. 5, No. 3, 1967, pp. 6–18.

[34]. D. Singer, R. M. Golinkoff, K. Hirsh-Pasek (Eds.), Play=Learning: How play motivates and enhances children's cognitive and social-emotional growth, *Oxford University Press*, New York, 2006, pp. 74-100.

[35]. J. Huizinga, Homo Ludens: A Study of the Play Element in Culture, *Beacon Press*, Boston, 1955.

[36]. Sutton-Smith B., The Ambiguity of Play, *Harvard University Press*, Cambridge, MA, 1997.

[37]. Rodriguez H., The Playful and the Serious: An approximation to Huizinga's Homo Ludens, *Game Studies*, Vol. 6, Issue 1, December 2006.

[38]. L. H. Larsen, L. Schou, H. H. Lund, H. Langberg, The Physical Effect of Exergames in Healthy Elderly—A Systematic Review, *Games for Health*, Vol. 2, No. 4, 2013.

[39]. H. H. Lund, J. D. Jessen, Effects of Short-Term Training of Community-Dwelling Elderly with Modular Interactive Tiles, *Games for Health*, Vol. 3, No. 5, 2014, pp. 277-283.

[40]. G. Bateson, A Theory of Play and Fantasy, in The Game Design Reader: A Rules of Play Anthology, Salen Katie and Zimmermand Eric (Eds.), *The MIT Presse*, Cambridge, 2006.

[41]. G. Bateson, The message 'this is play, in Group processes: Transactions of the second conference, B. Schaffner (Ed.), *Josiah Macy, Jr. Foundation*, New York, 1956, pp. 145–242.

[42]. M. J. Apter, J. H. Kerr, A Structural Phenomenology of Play, in Adult Play: A reversal theory approach, John H. Kerr and Michael J. Apter (Eds.), *Swets and Zeitlinger*, Amsterdam, 1991.

[43]. C. Jessen, H. H. Lund, On play forces, play dynamics, and playware, Unpublished manuscript, Technical University of Denmark, 2008.

8.

Multiscale Modelling and Simulation of Fiber-Reinforced Plastics Under Impact Loading

Arash Ramezani and Hendrik Rothe

8.1. Introduction

This work will focus on composite armor structures consisting of several layers of ultra-high molecular weight polyethylene (UHMW-PE), a promising ballistic armor material due to its high specific strength and stiffness. The goal is to evaluate the ballistic efficiency of UHMW-PE composite with numerical simulations, promoting an effective development process. Due to the fact that all engineering simulation is based on geometry to represent the design, the target and all its components are simulated as CAD models. The work will also provide a brief overview of ballistic tests to offer some basic knowledge of the subject, serving as a basis for the comparison of the simulation results. Details of ballistic trials on composite armor systems are presented. Instead of running expensive trials, numerical simulations should identify vulnerabilities of structures. Contrary to the experimental result, numerical methods allow easy and comprehensive studying of all mechanical parameters. Modeling will also help to understand how the fiber-reinforced plastic armor schemes behave during impact and how the failure processes can be controlled to our advantage. By progressively changing the composition of several layers and the material thickness, the composite armor will be optimized. There is every reason to expect possible weight savings and a significant increase in protection, through the use of numerical techniques combined with a small number of physical experiments. After a brief introduction and description of the different methods of space discretization in Section III,

Arash Ramezani

Chair of Measurement and Information Technology, Institute of Automation Technology, University of the Federal Armed Forces Hamburg, Hamburg, Germany there is a short section on ballistic trials where the experimental set-up is depicted, followed by Section V describing the analysis with numerical simulations. The paper ends with a concluding paragraph in Section VI.

8.2. State-of-the-Art

The numerical modeling of composite materials under impact can be performed at a constituent level (i.e., explicit modeling of fibre and matrix elements, e.g., [1]), a meso-mechanical level (i.e., consolidated plies or fibre bundles, e.g., [2]), or macromechanically in which the composite laminate is represented as a continuum. In [3–6] a non-linear orthotropic continuum material model was developed and implemented in a commercial hydrocode (i.e., ANSYS® AUTODYN®) for application with aramid and carbon fibre composites under hypervelocity impact. The non-linear orthotropic material model includes orthotropic coupling of the material volumetric and deviatoric responses, a non-linear equation of state (EoS), orthotropic hardening, combined stress failure criteria and orthotropic energy-based softening. For more detail refer to [7]. Lässig, *et al.* [8] conducted extensive experimental characterization of Dyneema® HB26 UHMW-PE composite for application in the continuum non-linear orthotropic material model, and validated the derived material parameters through simulation of spherical projectile impacts at hypervelocity. The target geometry is homogenized. The projectile is an aluminum ball in simplified terms. However, homogenized target geometries with orthotropic material models are not able to reproduce different modes of failure. The results are valid for aluminum spherical-shaped projectiles in hypervelocity range only. Nguyen, *et al.* [9] evaluated and refined the modeling approach and material model parameter set developed in [8] for the simulation of impact events from 400 m/s to 6600 m/s. Across this velocity range the sensitivity of the numerical output is driven by different aspects of the material model, e.g., the strength model in the ballistic regime and the equation of state (EoS) in the hypervelocity regime. Here, the target geometry is divided into sub-laminates joined by bonded contacts breakable through a combined tensile and shear stress failure criterion. The models mentioned above are valid for blunt FSP´s from a velocity range of 400 to 6600 m/s. They show considerable shortcomings in simulating pointed projectiles and thick HB26-composites. This paper will present an optimal solution of this problem with an enhanced model for ultra-high molecular weight polyethylene under impact loading. For the first time, composite armor structures consisting of several layers of fiber-reinforced plastics are simulated for all the current military threats.

8.3. Methods of Space Discretization

To deal with problems involving the release of a large amount of energy over a very short period of time, e.g., explosions and impacts, there are three approaches: as the problems are highly non-linear and require information regarding material behavior at ultra-high loading rates which is generally not available, most of the work is experimental and thus may cause tremendous expenses. Analytical approaches are possible if the geometries involved are relatively simple and if the loading can be described through boundary conditions, initial conditions or a combination of the two. Numerical solutions are far more general in scope and remove any difficulties associated with geometry [10]. They apply an explicit method and use very small time steps for stable results. For problems of dynamic fluid-structure interaction and impact, there typically is no single best numerical method which is applicable to all parts of a problem. Techniques to couple types of numerical solvers in a single simulation can allow the use of the most appropriate solver for each domain of the problem. The goal of this paper is to evaluate a hydrocode, a computational tool for modeling the behavior of continuous media. In its purest sense, a hydrocode is a computer code for modeling fluid flow at all speeds [11]. For that reason a structure will be split into a number of small elements. The elements are connected through their nodes (see Fig. 8.1).

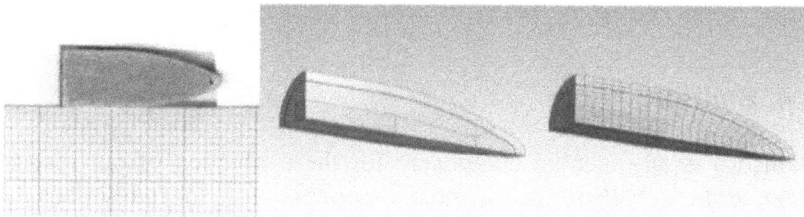

Fig. 8.1. Example grid.

The behavior (deflection) of the simple elements is well-known and may be calculated and analyzed using simple equations called shape functions. By applying coupling conditions between the elements at their nodes, the overall stiffness of the structure may be built up and the deflection/distortion of any node – and subsequently of the whole structure – can be calculated approximately [12]. Using a CAD-neutral environment that supports bidirectional, direct, and associative interfaces with CAD systems, the geometry can be optimized successively [13]. Therefore, several runs are necessary: from modeling to calculation to the evaluation and subsequent improvement of the model (see Fig. 8.2).

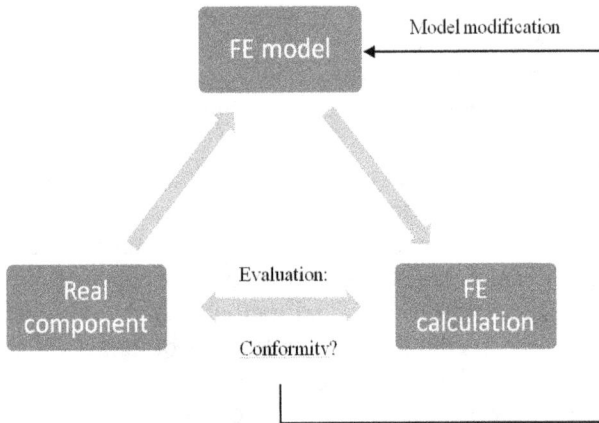

Fig. 8.2. Basically iterative procedure of a FE analysis [12].

Bullet-resistant materials are usually tested by using a gun to fire a projectile from a set distance into the material in a set pattern. Levels of protection (see Fig. 8.3) are based on the ability of the target to stop a specific type of projectile traveling at a specific speed.

8.4. Ballistic Trials

Ballistics is an essential component for the evaluation of our results. Here, terminal ballistics is the most important sub-field. It describes the interaction of a projectile with its target. Terminal ballistics is relevant for both small and large caliber projectiles. The task is to analyze and evaluate the impact and its various modes of action. This will provide information on the effect of the projectile and the extinction risk. Given that a projectile strikes a target, compressive waves propagate into both the projectile and the target. Relief waves propagate inward from the lateral free surfaces of the penetrator, cross at the centerline, and generate a high tensile stress. If the impact was normal, we would have a two-dimensional stress state. If the impact was oblique, bending stresses will be generated in the penetrator. When the compressive wave reached the free surface of the target, it would rebound as a tensile wave. The target may fracture at this point. The projectile may change direction if it perforates (usually towards the normal of the target surface).

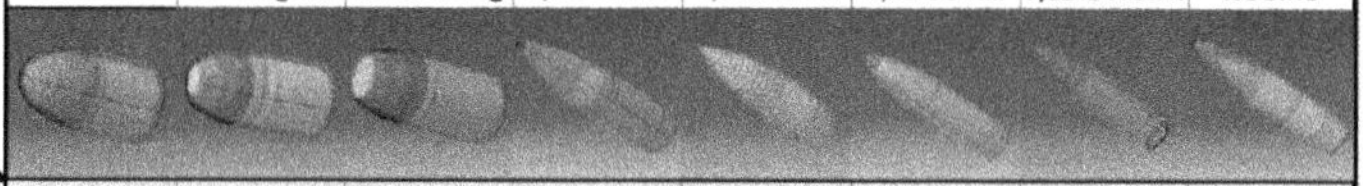

	Projectile	9 x 19 mm	.357 Magnum	.44 Rem. Mag.	5,56x45 mm	7,62x39 mm	7,62x51 mm	7,62x54 mm R	.50 BMG
Protection Level									
1	PM 1 / VR 1								
2	PM 2 / VR 2	$v = 360 \pm 10 \frac{m}{s}$ $E = 518\,J$							
3	PM 3 / VR 3	$v = 415 \pm 10 \frac{m}{s}$ $E = 689\,J$							
4	PM 4 / VR 4		$v = 430 \pm 10 \frac{m}{s}$ $E = 943\,J$	$v = 440 \pm 10 \frac{m}{s}$ $E = 1510\,J$					
5	PM 5 / VR 5		$v = 580 \pm 10 \frac{m}{s}$ $E = 1194\,J$						
6	PM 6 / VR 6					$v = 720 \pm 10 \frac{m}{s}$ $E = 2074\,J$			
7	PM 7 / VR 7 STANAG Level 1				$v = 950 \pm 10 \frac{m}{s}$ $E = 1805\,J$		$v = 830 \pm 10 \frac{m}{s}$ $E = 3289\,J$		
8	PM 8 / VR 8 STANAG Level 2					$v = 740 \pm 10 \frac{m}{s}$ $E = 2108\,J$			
9	PM 9 / VR 9						$v = 820 \pm 10 \frac{m}{s}$ $E = 3261\,J$		
10	PM 10 / VR 10 STANAG Level 3							$v = 860 \pm 10 \frac{m}{s}$ $E = 3846\,J$	
11	PM 11 STANAG Level 3						$v = 930 \pm 10 \frac{m}{s}$ $E = 3633\,J$		
12	PM 12						$v = 810 \pm 10 \frac{m}{s}$ $E = 4166\,J$		
13	PM 13								$v = 930 \pm 10 \frac{m}{s}$ $E = 10595\,J$

Fig. 8.3. Basically iterative procedure of a FE analysis [12].

Because of the differences in target behavior based on the proximity of the distal surface, we must categorize targets into four broad groups. A semi-infinite target is one where there is no influence of distal boundary on penetration. A thick target is one in which the boundary influences penetration after the projectile is some distance into the target. An intermediate thickness target is a target where the boundaries exert influence throughout the impact. Finally, a thin target is one in which stress or deformation gradients are negligible throughout the thickness. There are several methods by which a target will fail when subjected to an impact. The major variables are the target and penetrator material properties, the impact velocity, the projectile shape (especially the ogive), the geometry of the target supporting structure, and the dimensions of the projectile and target. In order to develop a numerical model, a ballistic test program is necessary. The ballistic trials are thoroughly documented and analyzed – even fragments must be collected. They provide information about the used armor and the projectile behavior after fire, which must be consistent with the simulation results (see Fig. 8.4). In order to create a data set for the numerical simulations, several experiments have to be performed. Ballistic tests are recorded with high-speed videos and analyzed afterwards. The experimental set-up is shown in Fig. 8.5. Testing was undertaken at an indoor ballistic testing facility. The target stand provides support behind the target on all four sides. Every ballistic test program includes several trials with different composites. The set-up has to remain unchanged.

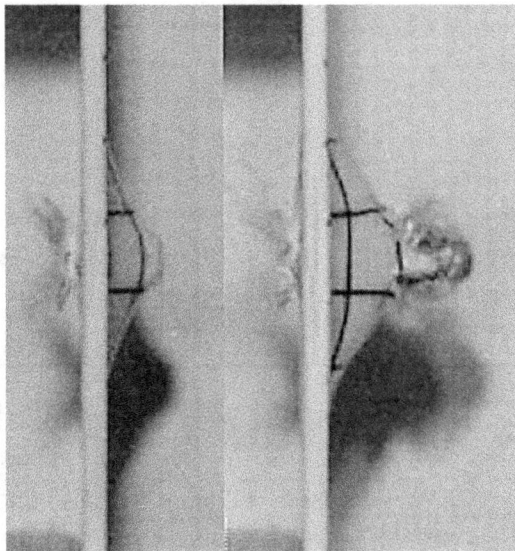

Fig. 8.4. Ballistic tests and the analysis of fragments.

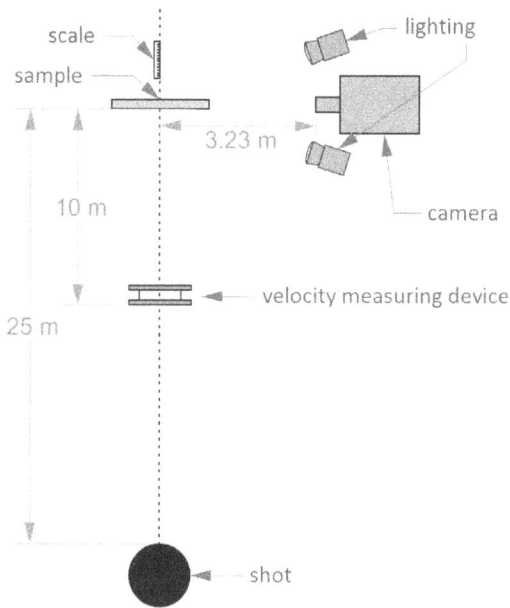

Fig. 8.5. Experimental set-up.

The camera system is a PHANTOM v1611 that enables fast image rates up to 646,000 frames per second (fps) at full resolution of 1280×800 pixels. The use of a polarizer and a neutral density filter is advisable, so that waves of some polarizations can be blocked while the light of a specific polarization can be passed.

Several targets of different laminate configurations were tested to assess the ballistic limit (V50). The ballistic limit is considered the velocity required for a particular projectile to reliably (at least 50 % of the time) penetrate a particular piece of material [15]. After the impact, the projectile is examined regarding any kind of change it might have undergone.

8.5. Numerical Simulation

The ballistic tests are followed by computational modeling of the experimental set-up. Then, the experiment is reproduced using numerical simulations. Fig. 8.1 shows a cross-section of the projectile and a CAD model. The geometry and observed response of the laminate to ballistic impact is approximately symmetric to the axis through the bullet impact point. Numerical simulation of modern armor structures requires the selection of appropriate material models for the constituent materials and the derivation of suitable material model input data. The laminate system studied here is an ultra-high molecular weight polyethylene composite. Lead and copper are also required for the projectiles.

The projectile was divided into different parts - the jacket and the base - which have different properties and even different meshes. These elements have quadratic shape functions and nodes between the element edges. In this way, the computational accuracy, as well as the quality of curved model shapes increases. Using the same mesh density, the application of parabolic elements leads to a higher accuracy compared to linear elements (1st order elements).

8.5.1. Modelling

In [8], numerical simulations of 15 kg/m^2 Dyneema® HB26 panels impacted by 6 mm diameter aluminum spheres between 2052 m/s to 6591 m/s were shown to provide very good agreement with experimental measurements of the panel ballistic limit and residual velocities, see Fig. 8.6. The modelling approach and material parameter set from [8] were applied to simulate impact experiments at velocities in the ballistic regime (here considered as < 1000 m/s). In Fig. 8.6 the results of modelling impact of 20 mm fragment simulating projectiles (FSPs) against 10 mm thick HB26 are shown. The model shows a significant under prediction of the ballistic limit, 236 m/s compared to 394 m/s.

8.5.2. Simulation Results

Relatively newer numerical discretization methods, such as Smoothed Particle Hydrodynamics (SPH), have been proposed that rectifies the issue of grid entanglement. The SPH method has shown good agreement with high velocity impact of metallic targets, better predictions of crack propagation in ceramics and fragmentation of composites under hypervelocity impact (HVI) compared to grid-based Lagrange and Euler methods. Although promising, SPH suffers from consistency and stability issues that lead to lower accuracy and instabilities under tensile perturbation. The latter makes it unsuitable for use with UHMW-PE composite under ballistic impact, because this material derives most of its resistance to penetration when it is loaded in tension. For these types of problems, the grid-based Lagrangian formulation still remains the most feasible for modeling UHMW-PE composite. 3D numerical simulations were performed of the full target and projectile, where both were meshed using 8-node hexahedral elements. The projectile was meshed with 9 elements across the diameter. The target is composed of sub-laminates that are one element thick, separated by a small gap to satisfy the master-slave contact algorithm (external gap in AUTODYN®) and bonded together as previously discussed. The mesh size of the target is approximately equal to the projectile at the impact site. The mesh was then graded towards the edge, increasing in coarseness to reduce the computational load of the model. Since UHMW-PE composite has a very low coefficient of friction, force fit clamping provides little restraint.

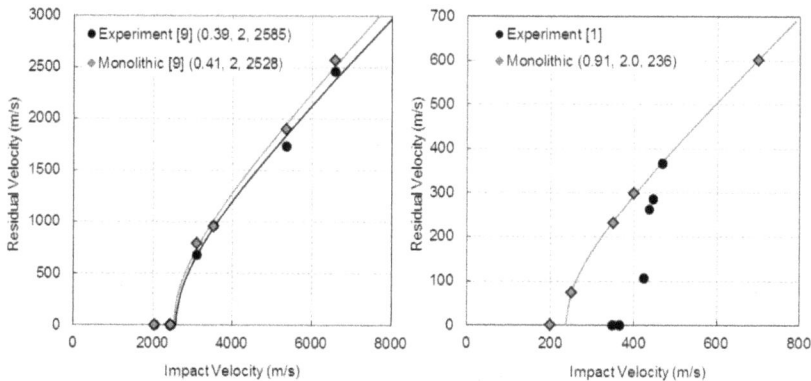

Fig. 8.6. Experimental and numerical impact residual velocity results for impact of 6 mm diameter aluminum spheres against 15 kg/m^2 Dyneema® HB26 at normal incidence (left) and impact of 20 mm fragment simulating projectiles against 10 mm thick Dyneema® HB26 at normal incidence (right). Lambert-Jonas parameters (a, p, Vbl) are provided in the legend.

High speed video of ballistic impact tests typical showed the action of loosening and moving clamps upon impact. As such no boundary conditions were imposed on the target. The FSP material was modelled as Steel S-7 from the AUTODYN® library described using a linear EoS and the Johnson-Cook strength model [16]. The aluminum sphere was modelled using AL1100-O from the AUTODYN® library that uses a shock EoS and the Steinburg Guinan strength model [17]. The master-slave contact algorithm was used to detect contact between the target and projectile.

The sub-laminate model with shock EoS was applied to the aluminum sphere hypervelocity impact series and 20 mm FSP ballistic impact series presented in Fig. 8.6, the results of which are shown in Fig. 8.7. The sub-laminate model is shown to provide a significant improvement in predicting the experimental V50 of 394 m/s for the FSP ballistic impacts (377 m/s) compared to the monolithic model (236 m/s).

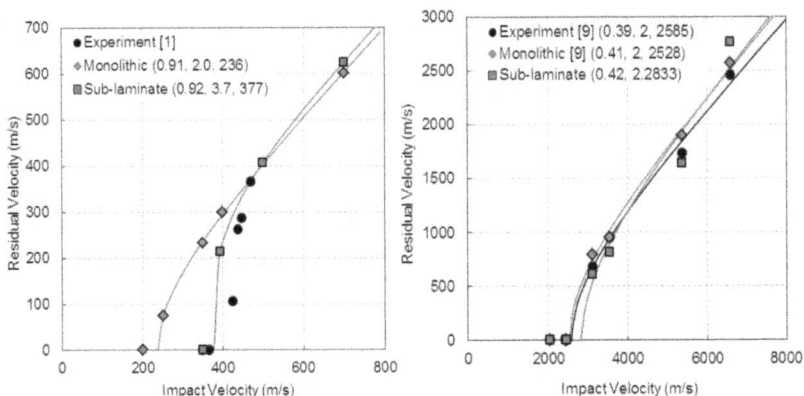

Fig. 8.7. Comparison of the experimental results with the two numerical models for impact of 20 mm fragment simulating projectiles against 10 mm thick Dyneema HB26® at normal incidence (left) , and impact of 6 mm diameter aluminium spheres against 15 kg/m² Dyneema® HB26 at normal incidence (right). Lambert-Jonas parameters (a, p, Vbl) are provided in the legend.

The ballistic limit and residual velocity predicted with the sub-laminate model for the hypervelocity impact case are shown to be comparable with the original monolithic model. For conditions closer to the ballistic limit, the sub-laminate model is shown to predict increased target resistance (i.e., lower residual velocity). For higher overmatch conditions there is some small variance between the two approaches. In Fig. 8.8, a qualitative assessment of the bulge formation is made for the 10 mm panel impacted at 365 m/s (i.e., below the V50) by a 20 mm FSP. Prediction of bulge development is important as it is characteristic of the material wave speed and is also a key measure in defence applications, particularly in personnel protection (i.e., vests and helmets). The sub-laminate model is shown to reproduce the characteristic pyramid bulge shape and drawing of material from the lateral edge. In comparison, the bulge prediction of the baseline model is poor, showing a conical shape with the projectile significantly behind the apex. In the baseline model penetration occurs through premature through-thickness shear failure around the projectile rather than in-plane tension (membrane) which would allow the formation of a pyramidal bulge as the composite is carried along with the projectile. Furthermore, in the baseline model the extremely small through thickness tensile strength (1.07 MPa) in the bulk material leads to early spallation/delamination of the back face. This allows the material on the target back face to fail and be accelerated ahead of the projectile. In the sub-laminate model, these two artifacts are addressed, and so a more representative bulge is formed.

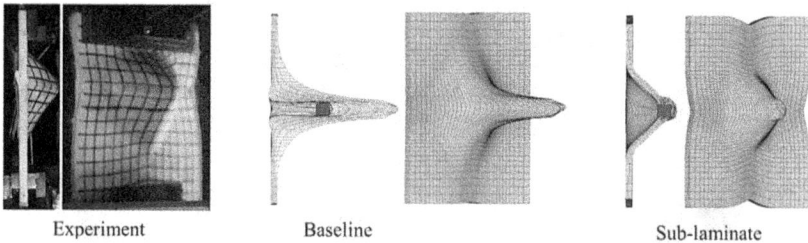

Experiment   Baseline   Sub-laminate

Fig. 8.8. Bulge of a 10 mm target impact by a 20 mm FSP at 365 m/s (experiment) and 350 m/s (simulations), 400 μs after the initial impact.

8.5.3. Further Validations

The material model developed in [8] and [9] has some shortcomings regarding the simulation of handgun projectiles (see Fig. 8.9). The ballistic limit was significantly under predicted. Evaluation of the result suggests that the failure mechanisms, which drive performance in the rear section of the target panel (i.e., membrane tension) were not adequately reproduced, suggesting an under-estimate of the material in-plane tensile performance.

A major difficulty in the numerical simulation of fibre composites under impact is the detection of failure processes between fibre and matrix elements as well as between the individual laminate layers (delamination). One promising approach is the use of "artificial" inhomogeneities on the macroscale. Here, an alternative simulation model has been developed to overcome these difficulties. Using sub-laminates and inhomogeneities on the macroscale, the model does not match the real microstructure, but allows a more realistic description of the failure processes mentioned above. Approaches based on the continuum or macroscale present a more practical alternative to solve typical engineering problems. However the complexity of the constitutive equations and characterization tests necessary to describe an anisotropic material at a macro or continuum level increases significantly.

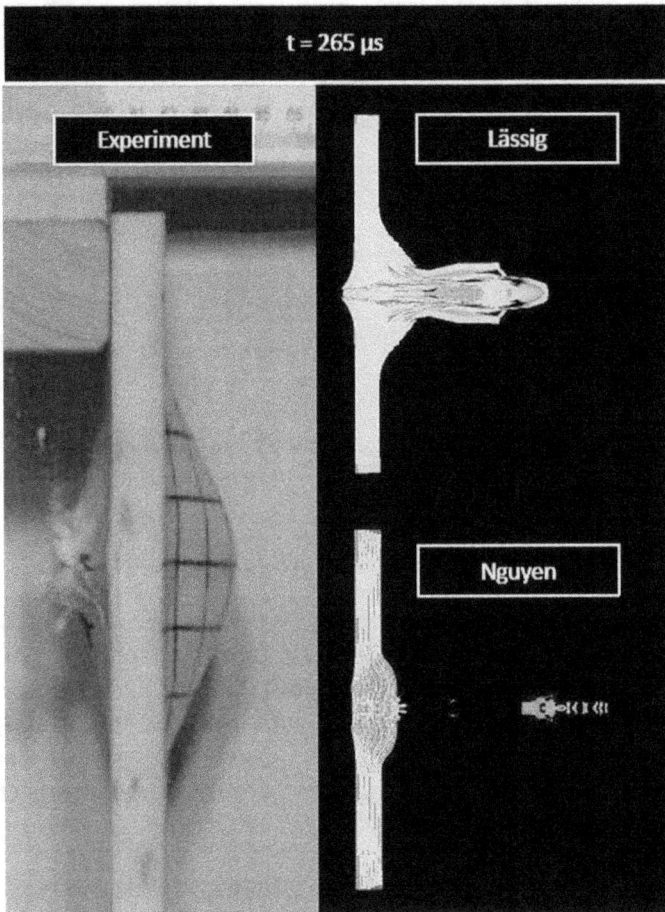

Fig. 8.9. Comparing experimental results with the previous simulation models of Lässig [8] and Nguyen [9], 265 µs after impact (grey = plastic deformation, green = elastic deformation, orange = material failure); projectile velocity: 674 m/s; target thickness: 16.2 mm (60 layers of HB26).

When considering the micromechanical properties of the orthotropic yield surface with a non-linear hardening description, a non-linear shock equation of state, and a three-dimensional failure criterion supplemented by a linear orthotropic softening description should be taken into account. It is important to consider all relevant mechanisms that occur during ballistic impact, as the quality of the numerical prediction capability strongly depends on a physically accurate description of contributing energy dissipation mechanisms. Therefore, a combination of ballistic experiments and numerical simulations is required. Predictive numerical tools can be extremely useful for enhancing our understanding of ballistic impact events. Models that are able to capture the key mechanical and thermodynamic processes can significantly improve our understanding of the phenomena by allowing time-resolved investigations of virtually every aspect of the impact event. Such high fidelity is immensely difficult, prohibitively expensive or near impossible to achieve with existing experimental measurement techniques. The thermodynamic response of a material and its ability to carry tensile and shear loads (strength) is typically treated separately within hydrocodes such that the stress tensor can be decomposed into volumetric and deviatoric components. Since the mechanical properties of fibre-reinforced composites are anisotropic (at least at the meso- and macroscale level), the deviatoric and hydrostatic components are coupled. That is deviatoric strains will produce a volumetric dilation and hydrostatic pressure leads to non-uniform strains in the three principal directions. The strength and failure model was investigated by modeling single elements under normal and shear stresses. It was found that under through-thickness shear stress, the element would fail prematurely below the specified through-thickness shear failure stress. It was found that if the through-thickness tensile strength was increased, failure in through-thickness shear was delayed. This evaluation study shows the importance of the strength, failure and erosion models for predicting performance in the ballistic regime. Previous material models for fiber-reinforced plastics were adjusted and the concept has been extended to different calibers and projectile velocities. Composite armor plates between 5.5 and 16.2 mm were tested in several ballistic trials and high-speed videos were used to analyze the characteristics of the projectile – before and after the impact. The simulation results with the modified model are shown in Fig. 8.10. The deformation of the projectile, e.g., 7.62×39 mm, is in good agreement with the experimental observation. Both delamination and fragmentation can be seen in the numerical simulation. Compared to the homogeneous continuum model, fractures can be detected easily. Subsequently, the results of experiment and simulation in the case of perforation were compared with reference to the projectile residual velocity. Here, only minor differences were observed. It should be noted that an explicit modeling of the individual fibres is not an option, since the computational effort would go beyond the scope of modern server systems (see Fig. 8.11).

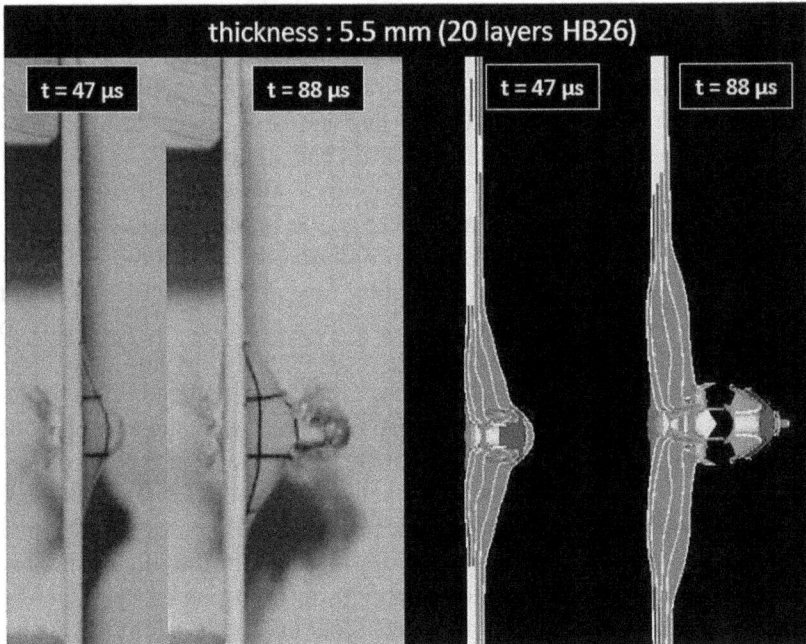

Fig. 8.10. Effect of a 5.5 mm target impact by a 7.62×39 mm bullet at 686 m/s, 47 μs and 88 μs after the initial impact.

8.6. Conclusions

This work demonstrated how a small number of well-defined experiments can be used to develop, calibrate, and validate solver technologies used for simulating the impact of projectiles on complex armor systems and composite laminate structures. Existing material models were optimized to reproduce ballistic tests. High-speed videos were used to analyze the characteristics of the projectile – before and after the impact. The simulation results demonstrate the successful use of the coupled multi-solver approach and new modeling techniques. The high level of correlation between the numerical results and the available experimental or observed data demonstrates that the coupled multi-solver approach is an accurate and effective analysis method.

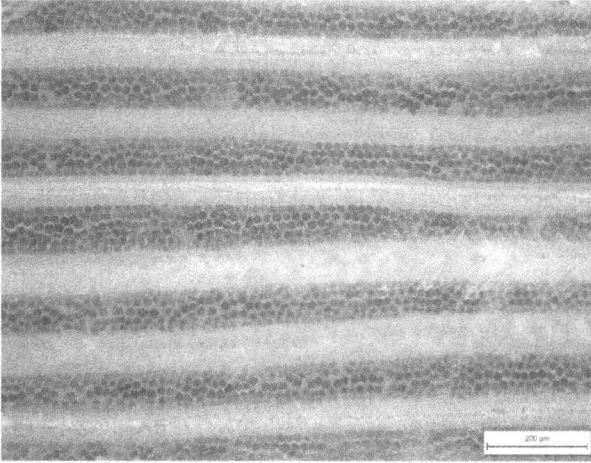

Fig. 8.11. Cross section of a Dyneema® HB26 panel.

A non-linear orthotropic continuum model was evaluated for UHMW-PE composite across a wide range of impact velocities. Although previously found to provide accurate results for hypervelocity impact of aluminum spheres, the existing model and dataset revealed a significant underestimation of the composite performance under impact conditions driven by through-thickness shear performance (ballistic impact of fragment simulating projectiles). The model was found to exhibit premature through thickness shear failure as a result of directional coupling in the modified Hashin-Tsai failure criterion and the large discrepancy between through-thickness tensile and shear strength of UHME-PE composite. As a result, premature damage and failure was initiated in the through-thickness shear direction leading to decreased ballistic performance. By de-coupling through-thickness tensile failure from the failure criteria and discretizing the laminate into a nominal number of kinematically joined sub-laminates through the thickness, progresses in modelling the ballistic response of the panels was improved. New concepts and models can be developed and easily tested with the help of modern hydrocodes. The initial design approach of the units and systems has to be as safe and optimal as possible. Therefore, most design concepts are analyzed on the computer. FEM-based simulations are well-suited for this purpose. Here, a numerical model has been developed, which is capable of predicting the ballistic performance of UHMW-PE armor systems. Thus, estimates based on experience are being more and more replaced by software. The gained experience is of prime importance for the development of modern armor. By applying the numerical model a large number of potential armor schemes can be evaluated and the understanding of the interaction between laminate components under ballistic impact can be improved. The most important steps during an FE analysis are the evaluation and interpretation of the outcomes followed by suitable modifications of the model. For that reason, ballistic trials are necessary to validate the simulation results. They are designed to obtain information about:

- The velocity and trajectory of the projectile prior to impact,

- Changes in configuration of projectile and target due to impact,

- Masses, velocities, and trajectories of fragments generated by the impact process. Ballistic trials can be used as the basis of an iterative optimization process. Numerical simulations are a valuable adjunct to the study of the behavior of metals subjected to high-velocity impact or intense impulsive loading. The combined use of computations, experiments and high-strain-rate material characterization has, in many cases, supplemented the data achievable by experiments alone at considerable savings in both cost and engineering man-hours.

References

[1]. D. B. Segala, P. V. Cavallaro, Numerical investigation of energy absorption mechanisms in unidirectional composites subjected to dynamic loading events, *Computational Materials Science*, Vol. 81, 2014, pp. 303–312.
[2]. S. Chocron, *et al.*, Modeling unidirectional composites by bundling fibers into strips with experimental determination of shear and compression properties at high pressures, *Composites Science and Technology*, Vol. 101, 2014, pp. 32–40.
[3]. C. J. Hayhurst, S. J. Hiermaier, R. A. Clegg, W. Riedel, M. Lambert, Development of material models for Nextel and kevlar-expoxy for high pressures and strain rates, *International Journal of Impact Engineering*, Vol. 23, 1999, pp. 365–376.

[4]. R. A. Clegg, D. M. White, W. Riedel, W. Harwick, Hypervelocity impact damage prediction in composites: Part I—material model and characterisation, *International Journal of Impact Engineering,* Vol. 33, 2006, pp. 190–200.

[5]. W. Riedel, H. Nahme, D. M. White, R. A. Clegg, Hypervelocity impact damage prediction in composites: Part II—experimental investigations and simulations, *International Journal of Impact Engineering,* Vol. 33, 2006, pp. 670–680.

[6]. M. Wicklein, S. Ryan, D. M. White, R. A. Clegg, Hypervelocity impact on CFRP: Testing, material modelling, and numerical simulation, *International Journal of Impact Engineering,* Vol. 35, No. 12, 2008, pp. 1861–1869.

[7]. ANSYS. AUTODYN Composite Modelling Release 15.0. (http://ansys.com/ 2016.07.08).

[8]. T. Lässig, *et al.*, A non-linear orthotropic hydrocode model for ultra-high molecular weight polyethylene in impact simulations, *International Journal of Impact Engineering,* Vol. 75, 2015, pp. 110–122.

[9]. L. H. Nguyen, *et al.*, Numerical Modelling of Ultra-High Molecular Weight Polyethylene Composite Under Impact Loading, *Procedia Engineering,* Vol. 103, 2015, pp. 436–443.

[10]. J. Zukas, Introduction to hydrocodes, *Elsevier Science,* Oxford, 2004.

[11]. G.-S. Collins, An Introduction to Hydrocode Modeling, *Applied Modelling and Computation Group,* Imperial College London, 2002.

[12]. P. Fröhlich, FEM Application Practice, *Vieweg Verlag,* 2005.

[13]. H.-B. Woyand, FEM with CATIA V5, *J. Schlembach Fachverlag,* 2007.

[14]. R. Frieß, General basis for ballistic material, construction and product testing, *Ballistic Day in Ulm,* 2008.

[15]. D. E. Carlucci, S. S. Jacobson, Ballistics: Theory and Design of guns and ammunition, *CRC Press,* 2008.

[16]. G. Johnson, W. Cook, A constitutive model and data for metals subjected to large strains, high strain rates and high temperatures, in *Proceedings of the 7th International Symposium on Ballistics,* 1983, pp. 541–547.

[17]. D. Steinberg, Equation of state and strength properties of selected materials, *Lawrence Livermore National Laboratory California,* Livermore, CA, 1996.

9.

How to Improve Driving Perception on an Advanced Dynamic Simulator While Cornering

Florian Savona, Emmanuelle Diaz, Anca Stratulat, Vincent Honnet, Philippe Vars, Stéphane Masfrand, Vincent Roussarie and Christophe Bourdin

9.1. Introduction

On dynamic driving simulators, motion perception is produced by stimulating the vestibular and somatosensory systems in addition to the visual system [1]. However, the intricacy of the multisensory stimulations undergone when driving a car makes the optimisation of motion-based simulators rather complex. For instance, it has already been shown that motion is overestimated when simulated at 1-to-1 rate [2-4]. Technical tricks used to avoid this overestimation include the scale factor, called motion gain, and/or a combination of tilt and translation, called tilt-coordination [3]. However, both the gain and the tilt-coordination needed to reproduce a positive or negative acceleration (e.g., starting the car or braking) are highly dependent on the level of acceleration being simulated [4-5].

For turning manoeuvres, the control of the simulator appears to be more complex than for longitudinal manoeuvres because, in addition to lateral acceleration, there are also the yaw and the roll motions of the car to be simulated. However, the main sensory information the driver depends on when manoeuvring is lateral acceleration. The driver actually controls the speed and the trajectory of the car to keep this acceleration in a comfortable range and to ensure a safety margin [6-7]. In most dynamic driving simulators, lateral acceleration is simulated using the tilt-coordination technique (lateral translation and lateral tilt). Moreover,

Florian Savona
Aix Marseille Univ, CNRS, ISM, Marseille, France

while cornering, the car is subject not only to a lateral linear acceleration, but also to rotational motions, such as yaw and roll. These motion components too are taken into account in driving simulation, and are highly dependent on steering behaviour while cornering. Therefore, Berthoz, et al. (2013) proposed that motion gains (for lateral and rotational acceleration) should be within the range 0.4-0.75 [8]. However, a limitation of their study is that the gain for linear translations, roll and yaw and their interactions were not systematically varied according to the level of acceleration.

Addressing this issue, the present study conducted on the PSA Group dynamic driving simulator SHERPA² focuses on cornering manoeuvres [9]. It systematically adjusts the motion gains for the three lateral motion components (lateral, yaw and roll motions) according to several levels of lateral acceleration. To evaluate the individual effects of the three parameters on driving behaviour, a slalom driving task was selected. This slalom driving task has already been used successfully by several international teams to study the realism of driving simulators [10-15]. Through subjective and objective variable analyses, we aimed to identify and quantify the major determinants of motion perception and driving performance while cornering, and to identify the best set of parameters for each level of acceleration simulated. Our specific aim was mapping motion gain set-ups, to improve driving simulation realism for a wider range of lateral accelerations. We hypothesised that the motion gains for the different parameters do not need to be linked, and that they could differ according to the level of lateral acceleration [11, 16]. The paper is structured as follows. Section 9.2 describes the experiment (participants, devices, scenario etc.). Section 9.3 presents the results of the study. Sections 9.4 and 9.5 discuss the findings of the study and their potential applications.

9.2. Methods

9.2.1. Participants

27 volunteers (2 women and 25 men), aged between 22 and 49 (mean age: 28) participated in the study. All volunteers were PSA Group employees and none had significant experience of the simulator (average dynamic driving simulator experience less than 1.5 hours). The participants did not know the experimental objective and all gave their

prior written consent, in conformity with local ethical committee requirements.

9.2.2. Experimental Devices

The experimental device used for this experiment was the SHERPA². SHERPA² is a dynamic driving simulator equipped with a hexapod and an X-Y platform. The cell placed on the hexapod contains a fully-equipped half-cab Citroen C1 (2 front adjustable seats, seat belts, steering wheel, pedals, gearbox, rearview mirror and side-view mirrors) where the driver sits. The motion limits of the hexapod are ±30 cm, ±26.5 cm and ±20 cm, on X, Y and Z respectively [9]. Rotational movements are limited to ±18 deg, ±18 deg and ±23 degrees, on pitch, roll and yaw respectively. The X-Y motion platform can reproduce linear movements of 10 and 5 metres. The maximum longitudinal and lateral acceleration is 5 m/s², and is actually produced by a combination of tilt and translation (herein, lateral tilt/translation is referred to as "lateral motion").

9.2.3. Experimental Scenario

The vehicle dynamics model (car dynamics and engine sound) selected for the present experiment was a Peugeot 208 1.4 HDi. The visual scene consisted of a straight two-lane road (road width: 8 m). Guardrails were placed at both sides of the road to delineate the maximum permitted excursion of the car. The slalom driving scenario consisted of a series of 8 posts situated a constant distance apart (varying for each level of acceleration). In addition, multiple mini-cones were used to represent the optimal sinusoidal pathway and help the subjects perform the task. The posts were alternately placed 0.9 m to the right and to the left of the road centreline (Fig. 9.1).

The velocity of the car was set at 70 km/h. Then, by adjusting the distance between the posts, various theoretical lateral accelerations were imposed. This yielded three different slalom scenarios leading to three theoretical lateral accelerations of 1, 2 and 4 m/s², corresponding to post spacings of 86.39, 61.09 and 43.19 metres, respectively. The equation enabling calculation of the theoretical lateral acceleration was borrowed from Grácio, Wentik and Païs (2011) [10]:

$$a_y = a\left(\frac{\pi}{d}v\right)^2,$$
(9.1)

where a (2 meters) is the sinusoidal trajectory amplitude, d (86.39, 61.09 or 43.19 meters) is the distance between the posts and v (70 km/h) is the car velocity.

Fig. 9.1. Visual environment of the slalom task. The blue posts identify the slalom to be performed, while the orange mini-cones illustrate the ideal trajectory. The trees positioned along the road, the dotted white lines and the texture on the road generate a rich optic flow.

9.2.4. Task

Drivers were asked to perform a slalom course on the dynamic driving simulator by following the mini-cone path, without touching any posts or leaving the road. The run was performed in cruise control at a constant speed of 70 km/h. Starting from 0 km/h, the driver accelerated until the constant speed of 70 km/h was reached just before beginning the slalom. The driver stopped the slalom course 200 m after the last post, at a position marked by a gantry and a chequered flag.

9.2.5. Experimental Design

For each level of lateral acceleration (1, 2 and 4 m/s²), the motion gains of the 3 lateral components (lateral motion, yaw and roll) were individually varied, leading to a total of 25 different conditions (see Table 9.1) obtained thanks to a Design of Experiments (DoE) approach, in that case a Central Composite design (CCD) with 3 factors [17]. CCD designs start with a factorial design with center points and add "star" points to estimate curvature in order to build a response surface.

Table 9.1. The 25 motion conditions tested. For each slalom scenario (1, 2 and 3, defined by inter-post distance), different gains in lateral motion, roll and yaw were applied.

	Lateral Motion Gain			Roll Angle Gain	Yaw Acceleration Gain
Slalom	1	2	3	1, 2 & 3	1, 2 & 3
Condition					
1	0.2	0.2	0.2	0.2	0.2
2	0.8	0.8	0.6	0.2	0.2
3	0.2	0.2	0.2	0.8	0.8
4	0.8	0.8	0.6	0.8	0.8
5	0.2	0.2	0.2	0.2	0.2
6	0.8	0.8	0.6	0.2	0.8
7	0.2	0.2	0.2	0.8	0.8
8	0.8	0.8	0.6	0.8	0.8
9	0	0	0	0.5	0.5
10	1	1	0.8	0.5	0.5
11	0.5	0.5	0.4	0	0.5
12	0.5	0.5	0.4	1	0.5
13	0.5	0.5	0.4	0.5	0
14	0.5	0.5	0.4	0.5	1
15	0.5	0.5	0.4	0.5	0.5
16	0.5	0.5	0.4	0.5	0.5
17	0.5	0.5	0.4	0.5	0.5
18	0.5	0.5	0.4	0.5	0.5
19	0.5	0.5	0.4	0.5	0.5
20	-1	-1	-1	-1	-1
21	0	0	0	0	0
22	1	1	0.8	1	1
23	1	1	0.8	0	0
24	0	0	0	1	0
25	0	0	0	0	1

The motion conditions differed in the gains applied to the three simulator motion components. Slaloms 1, 2 and 3 respectively correspond to acceleration levels of 1, 2 and 4 m/s². Condition 20 corresponds to the current SHERPA² configuration. Each participant performed 1 trial per condition (25) and per slalom (3), giving a total of 75 trials divided into two sessions to avoid fatigue. The trials were organised using a Williams latin square experimental design. A Williams design is a generalized Latin square that is also balanced for first order carryover effects. Motion gains were chosen taking into account the physical limitations of the simulator (position, speed, linear and angular acceleration).

The first session began with a simulator familiarisation phase (10 min of rural driving) and a slalom learning phase (one trial for each slalom scenario without simulator motion). This first session was followed by twenty-five trials on one slalom (constant level of acceleration). The second session, performed four hours later, consisted of another slalom familiarisation phase and the 50 remaining trials on the two other slaloms. The order of presentation of the three different slaloms was balanced over the total panel of participants. Furthermore, at the end of each trial, the participants answered two questions providing information on their subjective perception of the realism of the vehicle behaviour and the ease of the task. Two 11-point qualitative scales were used, ranging from 0 ("Not Realistic" or "Not Easy) to 10 ("Very Realistic" or "Very Easy"). In addition, motion sickness was monitored throughout the experiment via a Motion Sickness Susceptibility Questionnaire (MSSQ) [16].

9.2.6. Data Analysis

The subjective variables were assessed from participants' answers to the two questions posed at the end of each trial. Thus, for each slalom scenario, two subjective variables (realism of vehicle behaviour and ease of task) were measured.

The objective variables were assessed by analysing the driving behaviour of the participants. During the driving task, some dynamic variables were measured from the vehicle and simulator (e.g., lateral acceleration, steering wheel angle, lateral position). All these measurements were used to conduct an objective analysis of driver behaviour.

The Steering Wheel Reversal Rate (SWRR) was calculated from steering wheel angle. SWRR is a performance indicator that quantifies the amount of steering wheel correction, which indicates the effort required to accomplish a given task [13]. This metric measures the frequency of steering wheel reversals larger than a finite angle, or gap. The magnitude of this gap, the gap size, is thus a key parameter for this metric [18]. In the present study, the number of reversals per slalom was counted. The steering signal was filtered using a second-order low-pass Butterworth filter with a cutoff frequency depending on the slalom scenario: 0.6, 2 and 5 Hz for the 1, 2 and 4 m/s² acceleration levels respectively. The algorithm for detecting the reversal was extracted from "Reversal Rate 2" in Östlund's study (2005), and a difference greater than or equal to 2° (gap size) indicates one reversal [18].

Driving accuracy was quantified as lateral deviation from the reference trajectory (centre of the mini-cone path) and computed as Root Mean Squared Error (RMSE) of the vehicle path.

For each subjective and objective variable, Principal Component Analysis (PCA) was performed to determine whether there was consensus among subjects. If no consensus was found, an ascendant hierarchical classification was performed. Thereafter, the mean of each homogenous group was calculated. A quadratic model was subsequently executed, modelling each subjective and objective variable according to the different factors involved in the experimental design, for each slalom scenario. A response surface was constructed so as to graphically represent how a given variable evolves according to the different factors. The model contains simple, interaction and quadratic effects. A student t test enabled us to identify the significant coefficient of the model. The tables for the models only list the influential factors.

9.3. Results

9.3.1. Subjective Analysis

9.3.1.1. Simulator Sickness

During the experiment, four subjects experienced motion sickness and were unable to complete all the experimental conditions (Misery Score ≥ 6). Three of these participants experienced motion sickness during the slalom scenario with the highest lateral motion gains (Condition 10, 22 or 23 in Table 9.1). The remaining twenty-three subjects were able to complete the experiment without experiencing serious motion sickness (average Misery Score = 0.78 ± 1.2).

9.3.1.2. Realism of Vehicle Behaviour

PCA showed no consensus among participants, so the data was centred and hierarchical clustering performed to identify homogeneous groups of subjects. This analysis distinguished between 2 groups of participants (G1 and G2), whose experimental results were analysed separately. The models were significant for each slalom scenario. Table 9.2 presents the coefficients of the model as calculated from the answers regarding realism, by slalom and by participant group. Only simple, interaction and quadratic effects significantly influencing perceived realism are listed in this table.

For the first slalom (1 m/s²), the significant coefficients obtained with the Student t test identify Lateral Motion as an important factor determining realism for both groups of participants (Table 9.2), while Roll was only an important factor for the second Group. This means that modifying lateral motion and roll significantly modifies the driver's perception of realism (linearly for roll and non-linearly for lateral motion). Yaw motion does not appear to be an influential factor and thus does not appear in the column "Slalom 1" of Table 9.2.

Based on Group 1 (G1)'s answers, the experimental model estimates that the configuration perceived as most realistic has a lateral motion gain = 0.5, a roll motion gain = 1 and a yaw motion gain = 0.

Based on Group 2 (G2)'s answers, the configuration best producing an impression of realism has a lateral motion gain = 0.85, a roll motion gain = 1 and a yaw motion gain = 0. Given that yaw motion is not an influential factor for either group, it is fixed at zero for the calculation of the response surface. In any event, its value does not significantly modify the perceived realism.

Fig. 9.2 is a 3D representation of the experimental model of lateral and roll motion gains showing the realism of vehicle behaviour as perceived by Group 2 during the first slalom. The response surface shows that the degree of realism increases with the amplitude of the lateral motion gain, reaching a maximum at 0.85. It also shows the importance of roll motion, which gives the best result at a value of 1.

Table 9.2. Coefficients of the influential factors from the models for the two groups (G1 & G2), regarding realism of vehicle behaviour for the three slaloms. Only the factors (simple, interaction or quadratic effects) significantly ($p < 0.05$) influencing perceived realism are included. " ^{2} " means that the effect is quadratic and " - " specifies the interaction between two simple factors.

		Slalom 1		Slalom 2		Slalom 3	
		Group 1 (G1)	Group 2 (G2)	Group 1	Group 2	Group 1	Group 2
Coefficient Names		Coef	Coef	Coef	Coef	Coef	Coef
B0	Model Constant	7.76	6.89	7.1	6.69	6.8	6.7
B1	Lateral Motion	-0.37	1.17	-0.8	1.38	-1.1	0.8
B2	Roll		0.41			-0.1	
B3	Yaw					-0.1	
B1-1	Lateral Motion^2	-1.23	-4.5	-1.6	-2.6	-1.4	-2.6
B2-2	Roll^2					0.15	
B1-3	Lateral Motion-Yaw					0.22	-0.4
B2-3	Roll-Yaw					-0.3	

In the second slalom, the only significant factor for both groups was lateral motion ($p < 0.01$, see Table 9.2). Lateral motion gains of 0.4 and 0.7 respectively for G1 (Fig. 9.3) and G2 led to the strongest perceptions of realism.

In the third slalom, a difference was observed between the two groups. For G1, all three motions (lateral motion, roll and yaw) were significant factors ($p < 0.01$, see Table 9.2). The strongest perception of realism was achieved with lateral, roll and yaw motion gains of respectively 0.25, 1 and 0. For G2, only lateral motion gain was a significant factor ($p < 0.01$, see Table 9.2), and only at 0.5.

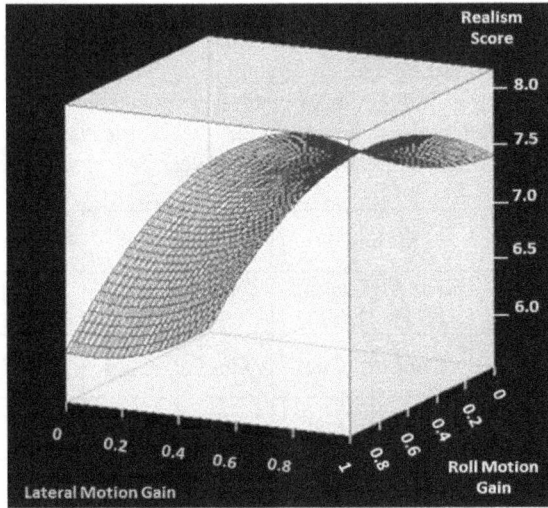

Fig. 9.2. 3D representation of the experimental model showing the realism of vehicle behaviour as perceived by Group 2 in the first slalom.

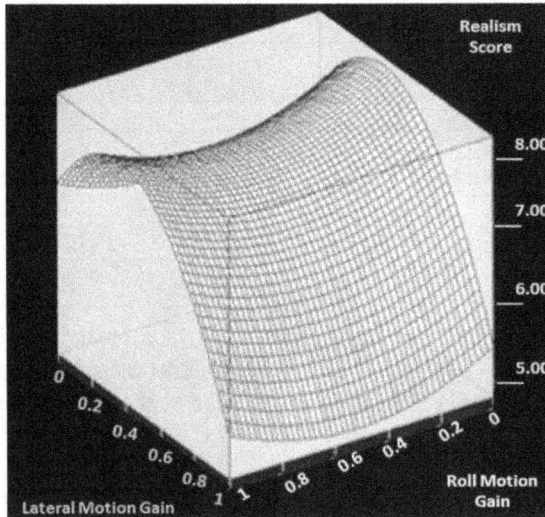

Fig. 9.3. 3D representation of the experimental model showing the realism of vehicle behaviour as perceived by Group 1 in the second slalom.

Moreover, the model highlights a significant interaction between lateral and yaw motions. This interaction is represented on the response surface in Fig. 9.4. If no lateral motion is produced, a yaw motion gain of 1 improves realism (red ellipse on Fig. 9.4). However, when there is lateral motion, realism is not improved by adding yaw, and can even be impaired e.g. white ellipse on Fig. 9.4.

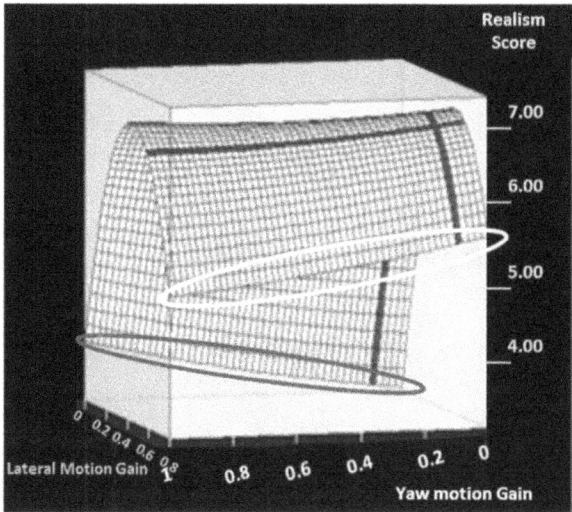

Fig. 9.4. 3D representation of the experimental model showing the realism of vehicle behaviour as perceived by Group 2 in the third slalom.

A comparison of the best configurations for lateral motion gain (only significant factor for all three slaloms) according to both groups and over the three slaloms is presented in Fig. 9.5. The lateral motion gains are degressive (decreasing with increased acceleration) for both groups.

Table 9.3 presents the best scores obtained for realism, for the two groups and the three slaloms. This table shows that a gain of 1 on roll motion always yields the best scores across slaloms.

9.3.1.3. Ease of the Task

Since the PCA revealed that there was consensus among participants for each slalom, the 23 participants' answers for the three slaloms were analysed together. The models were significant only for the second and the third slalom. Table 9.4 presents the model coefficients calculated on ease of the task for the second and third slaloms.

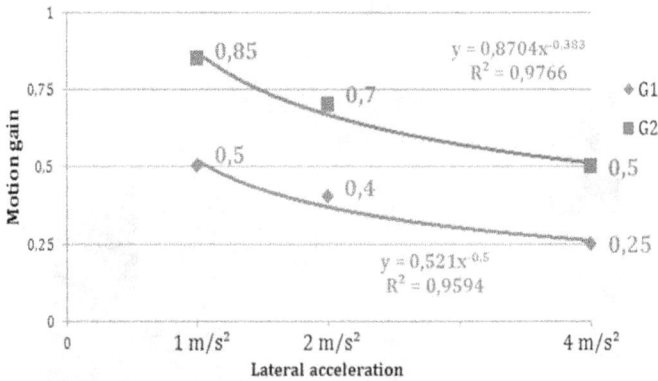

Fig. 9.5. Best lateral motion gains for the two groups and the three slaloms, obtained from model.

Table 9.3. Summary of motion gains perceived by the two groups of participants as yielding the best realism in the three slaloms. The two last columns show the realism score calculated by the model for each slalom scenario and group, with the associated roll and yaw motion gains (lateral motion gain is presented in the second column by group).

	Lateral Motion Gain G1/G2	Roll Motion Gain G1/G2	Yaw Motion Gain G1/G2	Best Score G1	Best Score G2
Slalom 1 **(1 m/s²)**	0.5/0.85	0 to 1/1	0 to 1/ 0 to 1	8.14 : Roll = 1 and Yaw = 0	8.23 : Roll =1 and Yaw = 0
Slalom 2 **(2 m/s²)**	0.4/0.7	0 to 1/ 0 to 1	0 to 1/ 0 to 1	7.97 : Roll = 0 and Yaw =0	8.45 : Roll = 1 and Yaw = 0
Slalom 3 **(4 m/s²)**	0.25/0.5	1/1	0/0	7.45 : Roll = 1 and Yaw = 0	7.06 : Roll = 1 and Yaw 0

For the first slalom, the absence of any difference between configurations in perception of the ease of the task can be explained by the fact that this slalom was very easy. Consequently, the participants probably did not need inertial information to perform the task, although the inertial information did not appear to make it harder for them (Table 9.4).

Table 9.4. Coefficients of influential factors from the models for all participants regarding ease of the task for the second and third slaloms. Only the factors (simple, interaction or quadratic effects) significantly ($p < 0.05$) influencing the ease of the task are included. " ² " means that the effect is quadratic.

		Slalom 2	Slalom 3
Coefficient Names		**Coef**	**Coef**
B0	Model Constant	8.23	6.97
B1	Lateral Motion	-0.96	-1.23
B1-1	Lateral Motion²	-0.71	-0.66

For the second and the third slalom, the only factor significantly modifying perception of the ease of the task was lateral motion ($p < 0.01$, see Table 9.4). Contrary to the first slalom, participants found the second and the third slalom substantially less easy to perform when the lateral motion gain was higher than 0.2 in the second slalom and 0 in the third (Fig. 9.6).

The motion gains creating the impression that the task was easier are presented in Table 9.5. The perceived ease of the task appears to depend both on the slalom's level of acceleration and on the lateral motion gain, which needs to be adjusted accordingly.

9.3.2. Objective Analysis

9.3.2.1. Steering Wheel Reversal Rate

Since the PCA revealed that there was consensus among participants for all the slaloms, all participants' scores were analysed together for each slalom. The models were significant for all three slaloms. Table 9.6 presents the model coefficients from the scores on reversal of the steering wheel for each the three slaloms.

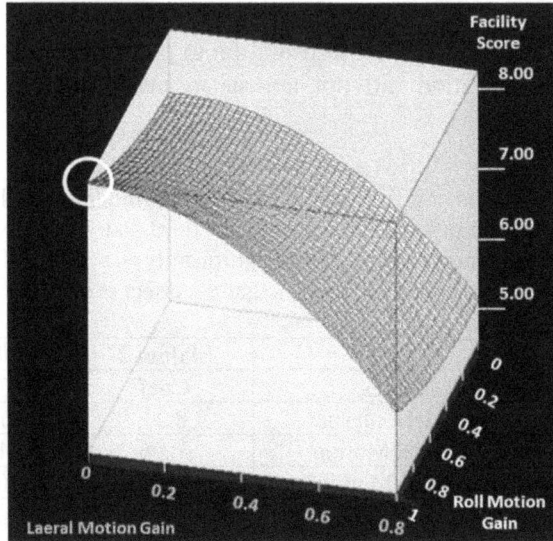

Fig. 9.6. 3D representation of the experimental model showing perceived ease of the task for the third slalom. The white circle represents the optimum of the model (Lateral motion gain = 0 and Roll = 1).

Table 9.5. Summary of motion gains perceived by all participants as rendering the task easiest in the three slaloms. The last column shows the ease score calculated by the model for each slalom, with the associated roll and yaw motion gains.

	Lateral Motion Gain	Roll Motion Gain	Yaw Motion Gain	Best Score
Slalom 1 (1 m/s²)	0 to 1	0 to 1	0 to 1	[8.4 - 9]
Slalom 2 (2 m/s²)	0.2	0 to 1	0 to 1	8.71: Roll = 0.3 and Yaw = 0
Slalom 3 (4 m/s²)	0	0 to 1	0 to 1	8.29: Roll = 1 and Yaw = 1

For the first slalom level, results show that the quantity of reversal decreases with increasing lateral motion. In other words, the lower the lateral motion gain, the greater the quantity of steering wheel reversal and the more steering wheel corrections are required to opmitise driving performance. The analysis also suggests that roll and yaw motion gains have no significant effect on driving performance, even though the best model is obtained with a roll and a yaw motion gain of 0 (Table 9.7).

Table 9.6. Coefficients of influential factors from the models for all participants regarding steering wheel reversal rate for the three slaloms. Only the factors (simple, interaction or quadratic effects) significantly ($p < 0.05$) influencing reversal of the steering wheel are included. " ^{2} " means that the effect is quadratic.

		Slalom 1	Slalom 2	Slalom 3
Coefficient Names		Coef	Coef	Coef
B0	Model Constant	9.1	10.38	10.16
B1	Lateral Motion	1.76	0.89	1.67
B1-1	Lateral Motion^2	1.14	3.4	2.58

Table 9.7. Summary of motion gains leading to lowest steering wheel reversal rates, for all participants in the three slaloms. The last column shows the SWRR score calculated by the model for each slalom, with the associated roll and yaw motion gains.

	Lateral Motion Gain	Roll Motion Gain	Yaw Motion Gain	Best Score SWRR
Slalom 1 (1 m/s^2)	1	0 to 1	0 to 1	7.87 : Roll and Yaw = 0
Slalom 2 (2 m/s^2)	0.5	0 to 1	0 to 1	9.86 : Roll and Yaw = 1
Slalom 3 (4 m/s^2)	0.25	0 to 1	0 to 1	9.2 : Roll and Yaw = 1

Contrary to the first slalom, the best models for the second and the third slalom were obtained with roll and yaw motion gains of 1, although these factors were not significant (see Table 9.7). For the second slalom, the best lateral motion gain is 0.5. Moreover, the lateral motion quadratic effect is significant in the model (see Table 9.6), as clearly visible on Fig. 9.7, where the lowest steering wheel reversal rate is obtained for a gain of 0.5 (white circle on Fig. 9.7). However at both ends of the lateral motion gain range (Gain = 0 and Gain = 1), there is much more steering wheel reversal.

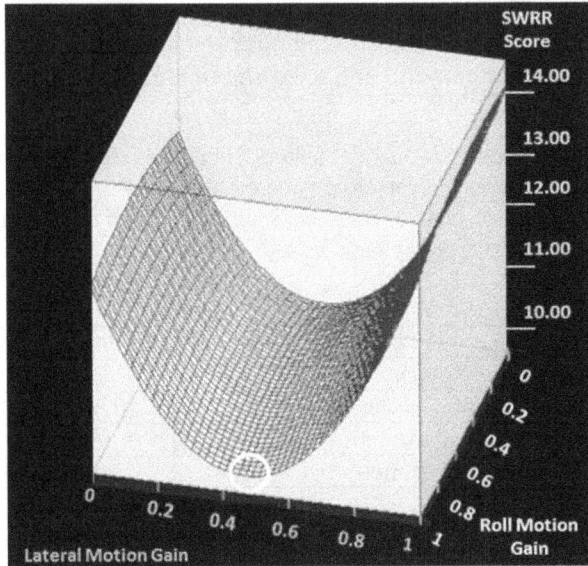

Fig. 9.7. 3D representation of the experimental model showing the steering wheel reversal rate (SWRR) for the second slalom. The white circle represents the optimum of the model (Lateral motion gain = 0.5 and Roll = 1).

Once again for the third slalom, the best SWRR reduction occurs with a lateral motion gain lower than in the second slalom (Lateral motion gain = 0.25). These results show that the lateral motion gain needs to be reduced when increased lateral acceleration is simulated, to see any significant improvement in driving skills.

9.3.2.2. Lateral Deviation from the Reference Trajectory

As with the previous variable, one group of participants was selected for the construction of the model. Table 9.8 presents the model coefficients regarding the root mean squared error (RMSE) of the lateral deviation for the third slalom. No difference was found among the motion configurations in the first and the second slalom (see Table 9.9). It is possible that the mini-cone path acted as an aid to accurate driving. Nevertheless, some differences were found in the third slalom (4 m/s²), the most difficult slalom and the one that subjects the driver to the greatest accelerations. Once again, the significant factor is lateral motion gain (Table 9.8).

Table 9.8. Coefficients of the influential factors from the models for all participants regarding RMSE of lateral deviation from the reference trajectory for the third slalom. Only the factors (simple, interaction or quadratic effects) significantly ($p < 0.05$) influencing the lateral deviation of the car are included. " ² " means that the effect is quadratic.

		Slalom 3
	Coefficient Names	**Coef**
B0	Model Constant	0.29
B1	Lateral Motion	0.03
B1-1	Lateral Motion²	0.06

For the third slalom, the experimental model identified two motion configurations leading to the same performance (see Table 9.9).

Table 9.9. Summary of motion gains leading to optimised car trajectory in the three slaloms and for all participants. The last column shows the RMSE score calculated by the model for each slalom, with the associated roll and yaw motion gains (the lateral motion gain is shown in the second column).

	Lateral Motion Gain	Roll Motion Gain	Yaw Motion Gain	**Best Score RMSE**
Slalom 1 (1 m/s²)	0 to 1	0 to 1	0 to 1	[0.177 – 0.204]
Slalom 2 (2 m/s²)	0 to 1	0 to 1	0 to 1	[0.228 – 0.254]
Slalom 3 (4 m/s²)				
First configuration	0.25	1	0	[0.275]
Second configuration	0.35	0 or 1	1	[0.275]

Fig. 9.8 shows the results of the third slalom for the second motion configuration (lateral motion gain of 0.35). It can be seen clearly that the curve is principally influenced by the magnitude of the lateral motion gain. When lateral motion gains are too low or too high, the participants widen their trajectories, leading to more errors and greater lateral deviation from the reference trajectory. Consequently, lateral motion gains at both ends of the range decrease driving accuracy.

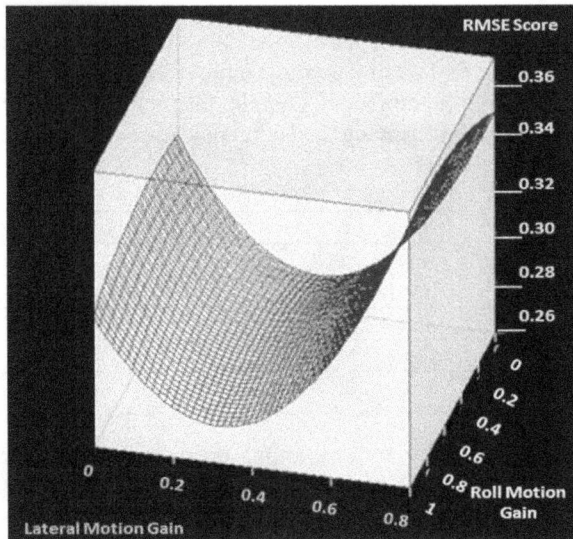

Fig. 9.8. 3D representation of the experimental model regarding RMSE of lateral deviation for the second slalom and the second configuration.

The results for the objective variables appear to corroborate those for the subjective variables concerning the evolution of lateral motion gain: the gain needs to decrease with increasing lateral acceleration. Nevertheless, it is difficult to identify a specific value to be recommended.

Although roll motion gain was never a significant factor in the objective variables (nor in the subjective perception of ease of performance), the best model results were obtained with a roll motion gain of 1, suggesting that this is the desirable value.

Yaw motion gain only emerged as an influential factor in one condition. In this case, the model advocates setting the yaw motion gain at 0 to optimise driving realism. However, for all other variables (subjective and objective), the models frequently calculated the best scores with a yaw motion gain of 1, notably for the slalom producing the greatest lateral acceleration and thus the highest yaw rate (Slalom 3).

9.4. Discussion of Results

The present study aims to explore motion gains on the three lateral motion components (lateral, roll and yaw motion) for different lateral acceleration levels. The slalom task used to assess their impact on driver performance has already been validated in the literature [11-14] on dynamic driving simulators. Previous studies recommended using the same motion gain, i.e. 0.6, for all three components to improve motion perception and driving behaviour [8, 15].

However, contrary to ours, these studies did not systematically modify motion gains according to different acceleration levels to simulate their effects on driving. Actually, they simply used the same motion gains for the three lateral components, without considering their individual effects on perceptions and driving performance. Our findings here suggest that the three lateral motion components should not be given the same gain, but that gains should be dynamically adjusted to the lateral acceleration level (i.e., non-constant gain). Consequently, this study contributes to the existing literature [8, 11] by taking into account several levels of lateral acceleration. These results are vital to a better understanding of the dynamic motion perception processes at play in a dynamic driving simulator. They offer the promise of finer simulation of lateral accelerations by applying different motion gains, thereby optimising perception processes and driving behaviour.

9.4.1. Motion Gains

9.4.1.1. Motion Realism

Results for the "Driving situation realism" variable suggest that there are two groups of drivers in the test population, distinguished mainly by their susceptibility to different lateral motion gains. The equations in Fig. 9.5, which characterise the amount of lateral motion as a function of lateral acceleration, can be used to parameterise the motion law in order to improve the perceptual validity of the simulator (realism). While the difference between the two groups remains unexplained (no driving factors common to a group were identified), results for both groups on this subjective variable emphasise the importance of using a gain that decreases with increased lateral acceleration. Although self-motion perception can tolerate significant discrepancies between the physical (inertial) and visual sources of the movement [19], the tilt-coordination technique was used to reproduce the lateral accelerations in the present study. Consequently, it is possible that tilt was more easily perceived when the lateral acceleration increased. Previous research showed that the threshold for lateral tilt (perceived as a tilt rather than a lateral acceleration) is higher for active drivers than for passive passengers [2, 20-21]. These studies [20-21] indicated that the maximum lateral tilt velocity was 6°/s, double that found for passive subjects [2]. In the present study, for the second and third slaloms and for the higher lateral motion gains, the simulator could reach 14° of inclination and an angular velocity of 12°/s (the limit set by the SHERPA[2] motion algorithm). This last magnitude is higher than those recommended by Nesti, et al. [21-22] and much higher than the inertial threshold of roll pitch (0.7°/s). Consequently, with these high lateral motion gains, it is very likely that the lateral inclination part of the lateral motion was not totally perceived as a sustained lateral acceleration, but rather as lateral rotation amplitude greater than the vehicle's natural roll [20-21, 23]. This may explain why participants assessed configurations with lateral motion gains significantly less than 1 as more realistic.

With regard to roll motion gain, the experimental model evaluated a gain of 1 as the most realistic situation.

Yaw motion gain, surprisingly, was never a factor positively influencing realism. However, an interaction with lateral motion was revealed for Group 2 in the third slalom. It appears that in the absence of lateral movement, the presence of the yaw (gain of 1) improves the realism of vehicle behaviour, as has already been shown in various studies [24-25].

This variable (realism) is very subjective and subject to interpretation. Thus, it is difficult to be sure that all participants had the same level of understanding or sense of presence. The multidimensional concept of presence is considered to be the ability of individuals to adopt a behaviour similar to that of everyday life and therefore their propensity to react to the various stimuli as if they were real [26]. Consequently, the sense of presence partially determines the general impression experienced by a driver in a driving simulator (particularly dynamic) and can vary from one individual to another [27-28]. Indeed, the only group differentiation observed is for this variable.

9.4.1.2. Ease of the Task and Objective Variables

9.4.1.2.1. Lateral Motion

Variables concerning the ease of the task, the number of steering wheel reversals and driving accuracy show that lateral motion gain needs to be reduced when lateral acceleration levels are increased, so as to improve self-motion perception and driving performance. This very important result is in line with findings from a previous study on the perception of longitudinal accelerations in a driving simulator, which showed that motion gain needed to evolve with acceleration level resulting from the braking of the vehicle [3]. Concerning the subjective perception of the ease of the task, a decrease in lateral motion gain was beneficial here when the slalom level increased. However, participants found it more difficult to perform the second and third slalom for configurations with lateral motion gains greater than 0.2 and 0 respectively. This can be explained by the fact that gains greater than these values cause greater body movements by the driver, leading to more discomfort and thus reduced driving ease.

In addition, objective analysis of steering corrections and lateral deviations also suggests that lateral motion gain needs to decrease with increasing slalom acceleration level. However, and contrary to the results on ease of the task, a gain of less than 0.2 is not recommended for accurate driving. Actually, varying the amount of lateral motion in a simulated slalom affects driver performance. Previous research showed a decrease in steering corrections when the lateral motion gain was increased [13]. In the present study, for the first slalom, a lateral motion gain of 1 produces optimal steering (minimum corrections to the steering wheel). This result supports the findings of Feenstra, et al. (2010), and even extends them. Feenstra, et al. (2010) also showed that cornering was more efficient when the lateral gain was 1. However, this result is only valid for a single lateral acceleration level (1.2 m/s²), and corresponds to the result obtained here for the first slalom level (for the SWRR variable). Nonetheless, driving performance in the two other slaloms is better with a lateral motion gain lower than 1 but greater than 0.2. From the analysis of both number of steering corrections and lateral driving accuracy (RMSE), a lateral motion gain of 0.25 for the third slalom can be recommended to optimise driving performance. Thus, when lateral acceleration levels are increased, and with them the difficulty of turning, the lateral motion gain should not remain fixed at 1, but should decrease progressively. It is surprising that multiplying the optimal gain for each slalom by that slalom's level of lateral acceleration always yields 1 m/s². Perhaps this surprising result is specific to slaloms, but a comparison with real driving would be required to confirm this. However, it appears that for a slalom task, the physical reproduction of lateral acceleration should be limited to 1 m/s² so as to optimise driving performance.

9.4.1.2.2. Roll Motion

A roll motion gain of 1 was perceived as the most realistic, leading to better driving performance, as confirmed by the results for the RMSE and SWRR variables. This finding represents a new advance in the field of simulation, and is consistent with results previously obtained with expert drivers [11]. Unlike Berthoz et al (2013), however, our study extends this finding to a wider population of non-expert driving subjects unaccustomed to virtual reality systems. Moreover, Berthoz, et al. (2013) did not test this roll motion gain in association with a lateral motion gain of less than 1. We were able to do so because the SHERPA² driving simulator used in the present study exactly reproduces the roll angle and its derivatives. In addition, the roll angle is particularly well synchronised with the visual roll. Although the usual threshold of inertial perception of roll motion is about 0.7°/s [21-22], in the presence of a visual stimulus the threshold rises to around 3°/s [21]. Thus, a roll motion with a gain of less than 1, as recommended up to now, may not provide the best results, often falling below the stimulation thresholds of the sensors. On the other hand, in the case of supra-threshold roll motion, this stimulation seems to reduce the latency of vections [29]. Thus, to improve realism and driving performance, it is particularly important that roll motion remains above the perception thresholds. A gain of 1 on this motion appears to have the desired effect.

9.4.1.2.3. Particular Case of Yaw Motion

The data from this experiment suggest that yaw motion is of little help when performing a slalom task. This contradicts fundamental studies showing that self-motion perception is better in presence of visuo-inertial stimuli than when stimuli are only visual [30]. Seeking an explanation for these surprising results, we thoroughly analysed all the parameters of the simulator. It emerged that the output data, i.e., those sent to the simulator by the motion cueing algorithm, were not consistent with the movements produced. In other words, it was subsequently found that the yaw motion produced by the SHERPA[2] simulator was well below that expected on the basis of the command sent. In particular, for a required yaw acceleration gain of 1, the output gain was between 0.3 and 0.5. This means that the yaw motion gains actually produced during the slalom tasks were greatly reduced, triggering infra-liminal stimulations and thus not, or only poorly, perceived by participants. Only slalom 3, with the strongest acceleration simulated at a gain of 1, produced yaw motion gains higher than the thresholds of inertial perception. Thus, the simulated yaw velocity only reached 2-3°/s, instead of the theoretical 12°/s generated by turning. Although the SHERPA[2] is capable of reproducing very high accelerations and yaw rates (600°/s² and 30°/s), the hexapod which allows this movement is also used to reproduce rapidly moving translational motions, as well as lateral tilt and roll. When all these movements are combined, yaw represents a very small proportion, which could explain these weak stimulations. In addition, the cut-off frequency used for the yaw acceleration filter may have been too high. This would mean that only a small part of the yaw information normally expected in these slaloms would be reproduced on the simulator. Consequently, we suggest that the absence of significant effect of yaw motion gain on the different variables studied can be explained by these physical limitations of the simulator. Moreover, the results of a previous study [31] indicate that a yaw rate between 5 and 6°/s (influence threshold) is required in the presence of visual information to significantly influence the perception of curvilinear trajectories. Here, although yaw velocity reached 2-3°/s in some slaloms, it is therefore very unlikely that this motion was sufficient to influence participants' behaviour.

There is a last, and non-exclusive, explanation for this lack of yaw motion effect. While our results appear to be in contradiction with a previous study [31], in that study poor visual stimulations were produced in association with one-dimensional (yaw) inertial stimulations. Here, however, participants experienced much richer stimulations. As a result, the presence of other inertial stimuli (lateral acceleration and roll) may mask the perception of yaw motion, thus pushing back its perception threshold [32-33]. This type of result has already been demonstrated by previous studies in flight simulators, which showed that yaw motion affected perception of the pilot's movement when it was the only motion reproduced. However, in the presence of lateral motion, the yaw effect was completely eclipsed, even though the two motions were above perception thresholds (2°/s for the yaw and 0.3 m/s² for the lateral). Moreover, the physical variables linking yaw rate and lateral acceleration depend on the linear velocity of displacement and the radius of curvature of this trajectory. The yaw rate/lateral acceleration relationship remains linear as long as it is the radius of the trajectory of the vehicle that is modified. However, when it is the displacement velocity that changes, lateral acceleration evolves quadratically at the displacement velocity, and the yaw rate/lateral acceleration relationship also changes in this manner. Thus, it is possible that at a steady driving speed and a small radius of curvature, the drivers can more easily feel and use yaw motion to improve their driving. Further work is needed to elucidate this point, particularly in a task requiring greater angular velocity, a larger rotational angle and lower linear velocity.

9.5. Conclusion

In conclusion, the results of this study clearly show that the lateral motion gain needs to be adjusted according to the level of lateral acceleration to be simulated, to improve the perceptual and behavioural validity of the simulator. This should guide work with dynamic driving simulators, and represents a new advance in the field of simulation. Previous studies of simulators may have missed this factor because they did not investigate the probable (independent) evolution of gains with the level of lateral acceleration reproduced [8, 11, 13]. The present study shows that even in driving simulation, multi-sensory perception processes are not linear, but evolve with the dynamics of stimulation, as has already been shown in fundamental studies [34-36]. However, tilt, in terms of acceleration, velocity and angle [37], should be the topic of a special study. The objective would be to shed light on the exact perceptual relationship between quantity of tilt and quantity of lateral translation reproduced.

Moreover, and despite the fact that our results on yaw motion need to be qualified in terms of the physical limitations observed, roll and yaw motion are shown here to have little influence on driving perception and performance. Our findings suggest, however, that a gain of 1 is the best setting for roll. There is not full agreement between our results for the subjective and the objective variables concerning the yaw effect. Yaw motion gain never seems to be a significant factor positively influencing the results in the present study, with its particular slalom task and lateral motions. However, it would be interesting to study yaw motion gain under different conditions, such as negotiating turns with small radii of curvature and with low longitudinal velocity [25]. Thus, given the present state of knowledge and the limitations regarding yaw motion with the SHERPA² simulator, it is difficult to recommend the use of a specific yaw motion gain. Until further investigations are available, therefore, the simulator motion cueing algorithm should be modified by decreasing the gain in lateral motion (decreasing gain), while keeping the roll motion gain to 1, to improve perceptions of realism and control performance.

References

[1]. A. Kemeny, F. Panerai, Evaluating perception in driving simulation experiments, *Trends in Cognitive Sciences*, Vol. 7, No. 1, January 2003, pp. 31-37.

[2]. E. L. Groen, W. Bles, How to use body tilt for the simulation of linear self motion, *J. Vestib. Res. Equilib. Orientat,* Vol. 14, No. 5, 2004, pp. 375-385.

[3]. A. Stratulat, V. Roussarie, J.-L. Vercher, C. Bourdin, Improving the realism in motion-based driving simulators by adapting tilt-translation technique to human perception, *IEEE Virtual Reality Conference*, 2011, pp. 47-50.

[4]. A. M. Stratulat, V. Roussarie, J.-L. Vercher, C. Bourdin, Perception of longitudinal acceleration on dynamic driving simulator, in *Proceeding of the Driving Simulation Conference*, Paris, France, 2012, pp. 33-40.

[5]. D. R. Berger, J. Schulte-Pelkum, H. H. Bülthoff, Simulating believable forward accelerations on a stewart motion platform, *ACM Trans. Appl. Percept*, Vol. 7, No. 1, January 2010, pp. 1-27.

[6]. E. Felipe, F. Navin, Automobiles on Horizontal Curves: Experiments and Observations, *Transp. Res. Rec*, Vol. 1628, No. 1, January 1998, pp. 50-56.

[7]. G. Reymond, A. Kemeny, J. Droulez, A. Berthoz, Role of lateral acceleration in curve driving: driver model and experiments on a real vehicle and a driving simulator, *Human Factors*, Vol. 43, No. 3, 2001, pp. 483-495.

[8]. A. Berthoz, *et al.*, Motion Scaling for High-Performance Driving Simulators, *IEEE Transaction Human-Machine System*, Vol. 43, No. 3, May 2013, pp. 265-276.

[9]. Chapron Thomas, Colinot Jean-Pierre, The new PSA Peugeot-Citroën Advanced Driving Simulator Overall design and motion cue algorithm, in

Proceeding of the Driving Simulation Conference, North America, Iowa City, 2007, pp. 44-52.

[10]. B. J. C. Grácio, M. Wentink, A. R. Valente Pais, Driver Behavior Comparison Between Static and Dynamic Simulation for Advanced Driving Maneuvers, *Presence Teleoperators Virtual Environ*, Vol. 20, No. 2, April 2011, pp. 143-161.

[11]. M. Dagdelen, J.-C. Berlioux, F. Panerai, G. Reymond, A. Kemeny, Validation Process of the Ultimate high-performance driving simulator, in *Proceeding of the Driving Simulation Conference*, Paris, France, 2006, pp. 37-47.

[12]. P. Feenstra, M. Wentink, B. Correia Grácio, W. Bles, Effect of Simulator Motion Space on Realism in the Desdemona Simulator, in *Proceeding of the Driving Simulation Conference*, Monaco, Europe, 2009.

[13]. P. Feenstra, R. van der Horst, B. J. C. Grácio, M. Wentink, Effect of Simulator Motion Cuing on Steering Control Performance: Driving Simulator Study, *Transp. Res. Rec. J. Transp. Res. Board*, Vol. 2185, No. 1, December 2010, pp. 48-54.

[14]. N. Filliard, B. Vailleau, G. Reymond, A. Kemeny, Combined Scale Factors for Lateral and Yaw Motion Rendering, in *Proceeding of the Driving Simulation Conference*, Monaco, Europe, 2009.

[15]. P. Pretto, M. Ogier, H. H. Bülthoff, J.-P. Bresciani, Influence of the size of the field of view on motion perception, *Comput. Graph.*, Vol. 33, No. 2, 2009, pp. 139–146.

[16]. B. J. Correia Grácio, J. E. Bos, M. M. Paassen, M. Mulder, Perceptual scaling of visual and inertial cues: Effects of field of view, image size, depth cues, and degree of freedom, *Experimental Brain Research*, Vol. 232, No. 2, November, 2013, pp. 637-646.

[17]. W. Tinsson, The notion of experimental plan, in *Experimental Plans: Buildings and Statistical Analysis*, Berlin, Vol. 67, 2010, pp. 3-37.

[18]. J. Östlund, B. Peters, Driving Performance Assessment-Methods and Metrics, *Report AIDE IST-1-507674-IP (D 2.2.5)*, European Union, 2005.

[19]. A. R. Valente Pais, M. M. (René) Van Paassen, M. Mulder, M. Wentick, Perception Coherence Zones in Flight Simulation, *Journal of Aircraft*, Vol. 47, No. 6, November 2010, pp. 2039-2048.

[20]. A. Nesti, C. Masone, M. Barnett-Cowan, P. R. Giordano, H. H. Bülthoff, P. Pretto, Roll rate thresholds and perceived realism in driving simulation, in *Proceeding of the Driving Simulation Conference*, Paris, France, 2012, pp. 23-32.

[21]. A. Nesti, S. Nooij, M. Losert, H. H. Bülthoff, P. Pretto, Roll rate perceptual thresholds in active and passive curve driving simulation, *SAGE Journal SIMULATION*, Vol. 92, No. 5, 2016.

[22]. Y. Valko, R. F. Lewis, A. J. Priesol, D. M. Merfeld, Vestibular Labyrinth Contributions to Human Whole-Body Motion Discrimination, *Journal of Neuroscience*, Vol. 32, No. 39, 26 September 2012, pp. 13537-13542.

[23]. Z. Fang, F. Colombet, J.-C. Collinet, A. Kemeny, Roll Tilt Thresholds for 8 DOF Driving Simulators, in *Proceeding of the Driving Simulation Conference*, Paris, France, 2014, pp. 16.1-16.7.

[24]. G. P. Bertollini, Yi. Glaser, J. Szczerba, R. Wagner, The effect of motion cueing on simulator comfort, perceived realism and driver performance during low speed turning, in *Proceeding of the Driving Simulation Conference*, Paris, France, 2014.

[25]. J. H. Hogema, M. Wentink, G. P. Bertollini, Effects of Yaw Motion on Driving Behaviour, Comfort and Realism, in *Proceeding of the Driving Simulation Conference,* Paris, France, 2012, pp. 149-158.

[26]. M. Slater, B. Lotto, M. M. Arnold, M. V. Sanchez-Vives, How we experience immersive virtual environments: the concept of presence and its measurement, *Anuario de Psicologia*, Vol. 40, No.°2, 2009, pp. 193-210.

[27]. C. Deniaud, V. Honnet, B. Jeanne, D. Mestre, The concept of "presence" as a measure of ecological validity in driving simulators, *Journal of Interaction Science,* Vol. 3, No. 1, 2015.

[28]. G. Wallis, J. Tichon, J, Predicting the Efficacy of Simulator-based Training Using a Perceptual Judgment Task Versus Questionnaire-based Measures of Presence, *Presence Teleoperators Virtual Environments*, Vol. 22, No. 1, 2013, pp. 67–85.

[29]. E. L. Groen, I. P. Howard, B. S. Cheung, Influence of body roll on visually induced sensations of self-tilt and rotation, *Perception*, Vol. 28, No. 3, 1999, pp. 287-297.

[30]. F. Savona, A. M. Stratulat, V. Roussarie, C. Bourdin, The influence of yaw motion on the perception of active vs passive visual curvilinear displacement, *Journal of Vestibular Research*, Vol. 25, 2015, No. 3, 4, pp. 125-141.

[31]. F. Savona, La perception des accélérations latérales en simulateur de conduite: étude de l'intégration multi sensorielle pour l'amélioration des performances de simulation, *Aix-Marseille University*, 2016.

[32]. P. R. Grant, B. Yam, R. Hosman, J. A. Schroeder, Effect of Simulator Motion on Pilot Behavior and Perception, *Journal of Aircraft*, Vol. 43, No. 6, 2006, pp. 1914–1924.

[33]. E. Groen, H. Smaili, R. Hosman, Simulated Decrab Maneuver: Evaluation with a Pilot Perception Model, *American Institute of Aeronautics and Astronautics*, 2005.

[34]. D. E. Angelaki, Y. Gu, G. C. DeAngelis, Multisensory integration: psychophysics, neurophysiology, and computation, *Current Opinion in Neurobiology*, Vol. 19, No. 4, 2009, pp. 452–458.

[35]. R. Jürgens, W. Becker, Human spatial orientation in non-stationary environments: relation between self-turning perception and detection of surround motion, *Experimental Brain Research*, Vol. 215, 2011, pp. 327–344.

[36]. M. L. Morgan, G. C. Deangelis, D. E. Angelaki, Multisensory integration in macaque visual cortex depends on cue reliability, *Neuron*, Vol. 59, No. 4, 2008, pp. 662–673.

[37]. F. Colombet, Z. Fang, A. Kemeny, Pitch tilt rendering for an 8-DOF driving simulator, in *Proceeding of the Driving Simulation Conference & Exhibition*, Tübingen, Germany, 2015, pp. 55-62.

10.

Step Climbing Strategy for a Wheelchair

Hidetoshi Ikeda

10.1. Introduction

For human beings, the ability to move is vital to life. However, many have lost their ability to move due to accident or illness. A wheelchair can provide physically disabled people with the ability to move around, and most wheelchairs are simple and not too expensive. Furthermore, users can easily drive a wheelchair once they learn how to operate it. However, current wheelchairs are not perfect and need to be improved.

10.1.1. Wheelchair

Most wheelchairs consist of a chair and a wheel mechanism. The basic configuration of a wheelchair has not changed for a long time [1]. A wheelchair has two casters employed as front wheels and two individually driving rear wheels. A typical manual wheelchair is driven by the arms of the user. The right and left rear wheels each have hand rims (Fig. 10.1) so that the user can easily steer the vehicle by using the difference between the rotations of the right and left wheels (Fig. 10.2). The ease of operability is one of the advantages of wheelchairs.

Excellent energy efficiency is another advantage of wheelchairs. Users are able to move their bodies and their vehicles by using the force of their arms if they move short distances. In the case that a user cannot generate enough power to drive the wheelchair, an electric wheelchair is helpful.

However, because wheelchairs are not perfect, users encounter a lot of difficulties in their daily lives. For example, such difficulties include opening a door and reaching for an object on a high shelf (Figs. 10.3 and 10.4).

Hidetoshi Ikeda
National Institute of Technology, Department of Mechanical Engineering, Toyama College, Japan

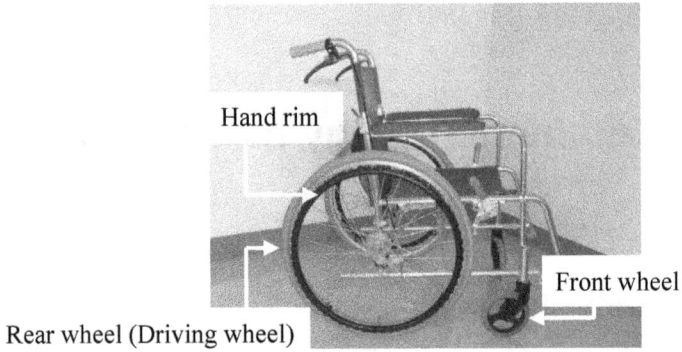

Fig. 10.1. Wheel arrangement of the typical manual wheelchair.

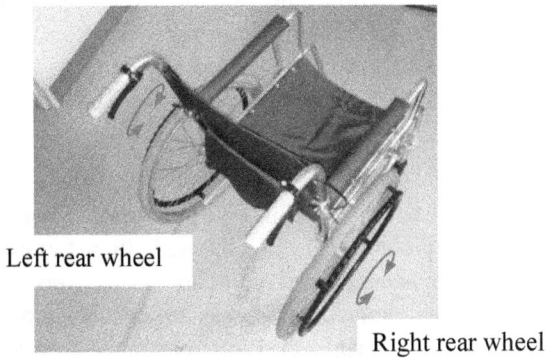

Fig. 10.2. Steering using two individually driving rear wheels.

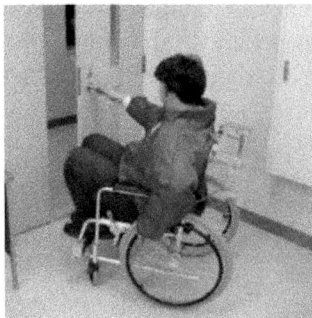

Fig. 10.3. Opening a door.

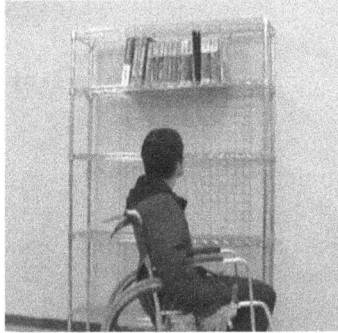

Fig. 10.4. Picking up objects.

Furthermore, wheelchairs have another difficulty. Some users have accidents when operating the wheelchair. Wheelchairs tend to tip over when they move on a slope or during step climbing. Some users fall from the chair when they transfer from the wheelchair to a toilet or bed. Calder and Kirby investigated the environmental factors and the accident ratios in wheelchair use [2], and they found that many people encounter problems when using wheelchairs.

In addition, the movement of the wheelchair is limited by rough terrain or even a low step, which is commonly found in the environment, and this is a huge problem for wheelchairs.

10.1.2. Related Research of Wheelchair Step Climbing

Without a human assistant, most wheelchair users are not able to enter an area that has steps. Thus, they need a special wheelchair equipped with a mechanism for step climbing or descending. Wheelchairs with such mechanisms have been widely researched and include, for example, a wheelchair with additional legs [3], a wheelchair with multiple wheels [4-5], a wheelchair with multiple wheels connected by active linkages [6], a wheelchair with an adjustable center of gravity [7], a wheelchair with a combination of an adjustable center of gravity and multiple wheels [8], a wheelchair with special wheels [9], and a tracked vehicle [10]. These mechanisms can provide the wheelchair with the ability to climb a step, stairs or surmount other obstacles. In addition, Mori [11] proposed a new step climbing method using a manual wheelchair equipped with linear actuator mechanisms and a light portable ramp.

The research group of the present report achieved cooperative step climbing of a wheelchair connected to a wheeled robot by passive links [12] and also step climbing and descending of a wheelchair and a wheeled robot with manipulators [13].

Other reports of multiple vehicles cooperating for crossing irregular terrain include those of Asama [14], who considered a forklift system. However, that research focuses on using wheeled robots and not wheelchairs.

10.1.3. Purpose of This Chapter

Some sections of this chapter are an extension of work originally presented in the *Second International Conference on Intelligent System and Applications* [13] and *Journal of the Robotics Society of Japan* (in Japanese) [15]. The sections of this chapter are organized as follows. Section 10.2 describes the analysis method of step climbing for a wheelchair, Section 10.3 describes the cooperative step climbing system and method of using a wheelchair and a robot. Section 10.4 provides a theoretical analysis of cooperative step climbing. Section 10.5 presents the experimental results, and Section 10.6 is the conclusion. The aim of the present chapter is to show wheelchair step climbing tactics and the theoretical analysis method.

10.2. Theoretical Analysis of Step Climbing for a Wheelchair

The purpose of this section is to show the theoretical analysis method of wheelchair step climbing.

A wheelchair has to fulfill some requirements in order to climb a step. These requirements correspond to *(1)* through *(4)* below and are shown in Fig. 10.5. We assume that the wheelchair has one pair of front wheels (casters) and one pair of rear wheels (driving wheels), the tires of the wheelchair do not transform, the wheelchair user does not change his posture in step climbing, the wheelchair is able to generate force to climb the step, and the wheelchair moves at a slow speed and keeps its balance in step climbing. These are analyzed by considering the system statics.

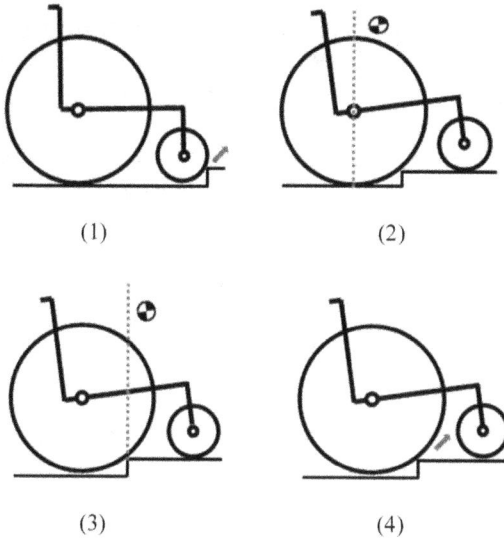

(1) (2)

(3) (4)

Fig. 10.5. Requirements to climb a step, *(1)* Exerting driving force on the ground to lift the front wheels, *(2)* Avoidance from tipping over backward in front wheel climbing, *(3)* Avoidance from tipping over backward in rear wheel climbing, *(4)* Exerting driving force on the step edge to lift up the rear wheels.

(1) When the front wheels (casters) begin to climb a step, the rear wheels (driving wheels) exert a driving force on the ground and do not slip. Here, the inclination of the wheelchair is zero.

(2) When the front wheels climb a step, the wheelchair does not tip over backward. The wheelchair center of gravity has to be in front of the contact point between the rear wheels and the ground after the front wheels lift up on the step. Here, the inclination of the wheelchair is determined by the step height.

(3) When the rear wheels begin to climb a step, the wheelchair does not tip over backward. The wheelchair center of gravity has to be in front of the contact point between the rear wheels and the step edge when the rear wheels leave the ground level (red region in diagram below).

step

(4) When the rear wheels begin to climb a step, the wheels exert a driving force on the step edge and do not slip. The inclination of the wheelchair is the angle when the rear wheels leave the ground level.

Requirements *(1)* and *(2)* exert the driving power to lift the front or rear wheels from the ground level. Requirements *(2)* and *(3)* prevent the wheelchair from tipping over backward.

Requirement *(3)* is more difficult than *(2)* to achieve. Thus, if the wheelchair achieves requirements *(1)*, *(3)* and *(4)*, it is able to get over the step.

10.2.1. Generating the Driving Force to Lift the Front Wheels (Requirement (1))

Fig. 10.6 shows the model of a wheelchair during step climbing of the front wheels, where M: mass of the wheelchair (wheelchair + driver), ρ: inclination of the wheelchair, h_m: height of the gravity center from the axes of the rear wheels, l: wheelbase, l_r: horizontal distance of the center of gravity from the rear axle, r: radius of the front wheels, R: radius of the rear wheels, f: driving force of the rear wheels, N_r: normal reaction affecting the rear wheels, N_f: normal reaction affecting the front wheels from the step edge, α: angle formed by the vertical line and the line from the step edge to the front wheel axle, g: gravitational acceleration, h: step height.

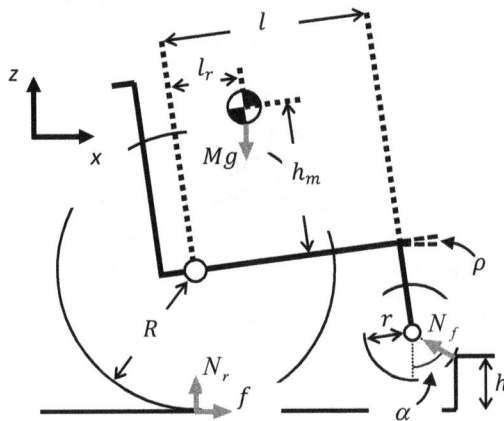

Fig. 10.6. Model of wheelchair during step climbing of front wheels.

We assume that the wheelchair (NOVA Integral-ME) is available in the market. The parameters of the wheelchair are $r = 0.063$ [m], $R = 0.3$ [m], $l = 0.43$ [m], $l_r = 0.149$ [m], $h_m = 0.371$ [m].

Based on preliminary measurements of the friction coefficients of a wheelchair in wet and dry conditions on asphalt, concrete, wood, and interior flooring, the ground surface considered in this chapter was assumed to have a friction coefficient in the range 0.6 to 0.9, which satisfies all of the above conditions.

When the wheelchair is moving in static equilibrium (Fig. 10.6), the equilibrium for both the x and z axes yields (10.1), (10.2).

$$f - N_f \sin \alpha = 0 \tag{10.1}$$

$$N_r - Mg + N_f \cos \alpha = 0 \tag{10.2}$$

The equilibrium of moments about the contact point between the rear wheels and the ground yields (10.3).

$$\{l \cdot \cos\rho + (R - r)\sin\rho + r \cdot \sin\alpha\} \cdot N_f \cos\alpha + h \cdot N_f \sin\alpha \\ - (l_r \cos\rho - h_m \sin\rho)Mg = 0 \tag{10.3}$$

From (10.3),

$$N_f = \frac{Mg \cdot l_r}{(l + r\sin\alpha)\cos\alpha + h\sin\alpha} \tag{10.4}$$

From (10.1) and (10.2),

$$f = N_f \sin\alpha \tag{10.5}$$

$$N_r = Mg - N_f \cos\alpha \tag{10.6}$$

The front wheels beginning to leave the road of the under step ($\rho = 0$) is the most difficult situation to climb the step. Fig. 10.7 shows the model of the front wheels when beginning to leave the ground level. In this case,

$$r - r\cos\alpha = h \tag{10.7}$$

Thus,

$$\cos\alpha = \frac{r-h}{r}, \qquad (10.8)$$

where

$$\sin\alpha = \sqrt{1-\cos^2\alpha} \qquad (10.9)$$

From (10.8), (10.9), we obtain

$$\sin\alpha = \frac{\sqrt{2rh-h^2}}{r} \qquad (10.10)$$

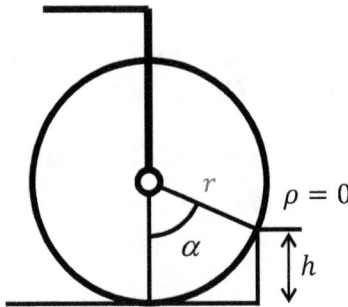

Fig. 10.7. Relation between r, h and α.

From (10.4)–(10.6),

$$\frac{f}{N_r} = \frac{l_r\sin\alpha}{\{(l+r\sin\alpha)\cos\alpha + h\sin\alpha\} - l_r\cos\alpha} \qquad (10.11)$$

(8), (10) are substituted for (11) and we obtain (12),

$$\frac{f}{N_r} = \frac{l_r\sqrt{2rh-h^2}}{(l-l_r)(r-h) + r\sqrt{2rh-h^2}} \qquad (10.12)$$

We define the proportional constant k_f ($0 \le k_f < 1$),

$$h = k_f \cdot r \qquad (10.13)$$

(13) is substituted for (12) and we obtain (14),

$$\frac{f}{N_r} = \frac{l_r \sqrt{k_f(2 - k_f)}}{(1 - l_r)(1 - k_f) + r\sqrt{k_f(2 - k_f)}} \qquad (10.14)$$

Using (10.14), we perform a numerical calculation to clarify the step climbing ability in front wheel climbing. The horizontal axis in Fig. 10.8 shows k_f, and the vertical axis shows f/N_r. The shading in Fig. 10.8 shows $f/N_r < \mu$, which indicates the front wheels are able to climb the step. Thus, the step height at which the wheelchair front wheels are able to climb is at most 56.4 % of the height of the front wheel radius when the friction coefficient is $\mu = 0.9$, and is just 23.6 % when the friction coefficient is $\mu = 0.5$.

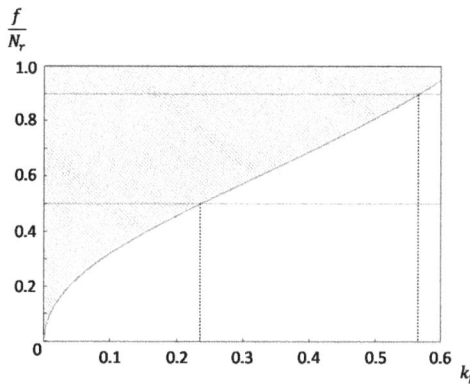

Fig. 10.8. Relation between step height and friction coefficient when the front wheels are able to climb the step.

10.2.2. Avoidance from Tipping over Backward (Requirements *(2)*, *(3)*)

As described above, requirement *(3)* is more difficult than *(2)* to achieve. Thus, in this section, we analyze requirement *(3)*.

Fig. 10.9 shows the model when the rear wheels begin to climb the step; here, β is the angle formed by the vertical line and the line from the step edge to the front wheel axle, and D_m is the horizontal distance between the center of gravity and the step edge. When the center of gravity is in front of the step edge, the wheelchair does not tip over backward. The requirement for avoiding tipping over backward in the climbing of the rear wheels is given in (10.15).

$$l_r \cos \rho - h_m \sin \rho - R \sin \beta \geq 0 \qquad (10.15)$$

When the rear wheels climb a step, the relation between R, β and h is calculated (Fig. 10.9 and (10.10)).

$$\sin \beta = \frac{\sqrt{2Rh - h^2}}{R} \qquad (10.16)$$

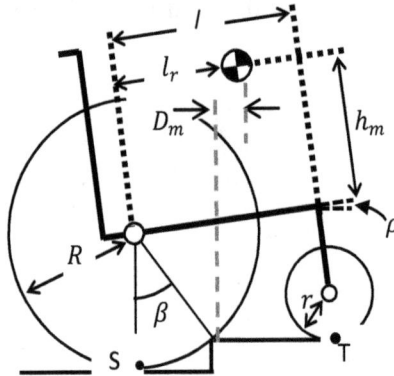

Fig. 10.9. Model of the rear wheels climbing while avoiding tipping over backward.

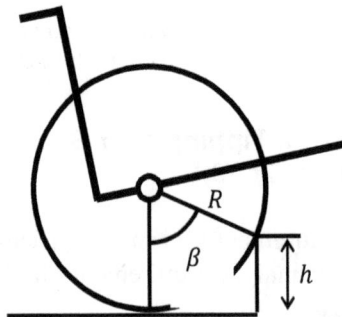

Fig. 10.10. Relation between *R, h* and *β*.

From (15), (16), we obtained the requirement to prevent tipping over backward,

$$l_r \cos\rho - h_m \sin\rho - \sqrt{2Rh - h^2} \geq 0, \qquad (10.17)$$

where S is the position of the bottom of the rear wheels, T is the contact point between the front wheels and the step (Fig. 10.9). When the rear wheels start to climb a step (leave the ground of the under step), the difference of heights between S and T is equal to step height, h. Thus,

$$h = R + l\sin\rho + (-R+r)\cos\rho - r \qquad (10.18)$$

and

$$\cos\rho = \sqrt{1 - \sin^2\rho} \qquad (10.19)$$

From (10.18), (10.19), we find

$$\sin\rho = \frac{l(h+r-R) - (-R+r)\sqrt{l^2 + h(2R-2r-h)}}{l^2 + (-R+r)^2} \qquad (10.20)$$

We define the proportional constant k_t $(0 \leq k_t < 1)$,

$$h = k_t \cdot r \qquad (10.21)$$

(10.21) is substituted for (10.20) and we obtain (10.22),

$$\sin\rho = \frac{l(k_t R + r - R) - (-R+r)\sqrt{l^2 + k_t R(2R-2r-k_t R)}}{l^2 + (-R+r)^2} \qquad (10.22)$$

Using (10.17), (10.19), (10.22), we perform a numerical calculation to clarify the limitation to avoiding tipping over backward.

The horizontal axis in Fig. 10.11 shows k_t and the vertical axis shows D_m. As shown, $D_m \geq 0$ indicates that the wheelchair is able to avoid tipping over backward. The tipping over backward of the wheelchair occurs when the wheelchair climbs a step whose height is over 9.16 % of the rear wheel radius.

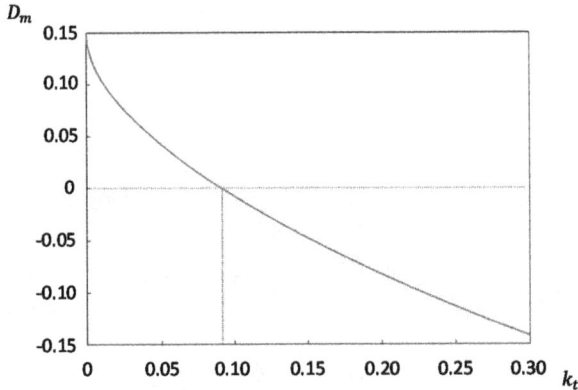

Fig. 10.11. Relation between step height and horizontal distance of center of gravity from step edge.

10.2.3. Generating the Driving Force to Lift the Rear Wheels (Requirement *(4)*)

Fig. 10.12 shows the model of a wheelchair during step climbing of the rear wheels. When the wheelchair is moving in static equilibrium, the equilibrium for both the x and z axes yields (10.23), (10.24).

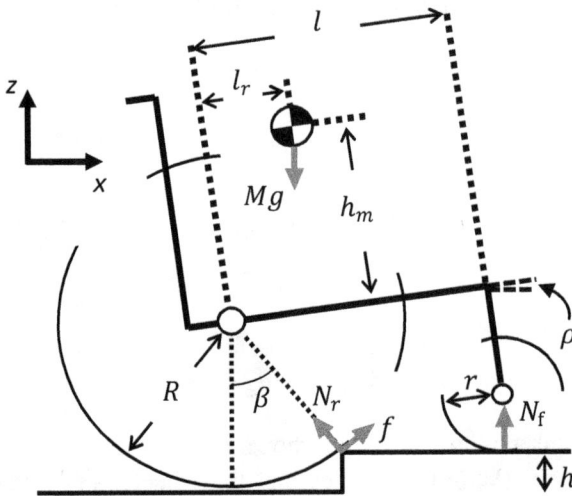

Fig. 10.12. Model of wheelchair during climbing of rear wheels.

$$f \cos \beta - N_r \sin \beta = 0 \qquad (10.23)$$

$$f \sin \beta + N_r \cos \beta + N_f - Mg = 0 \qquad (10.24)$$

From (10.23),

$$f \cos \beta \sin \beta - N_r \sin^2 \beta = 0 \qquad (10.25)$$

From (10.24),

$$f \sin\beta \cos\beta + N_r \cos^2 \beta + \left(N_f - Mg\right)\cos\beta = 0 \qquad (10.26)$$

Solving (10.25) – (10.24),

$$N_r = \left(Mg - N_f\right)\cos \beta \qquad (10.27)$$

From (10.23), (10.27)

$$f = \left(Mg - N_f\right)\sin \beta \qquad (10.28)$$

Thus, from (10.27), (10.28), we obtain

$$\frac{f}{N_r} = \tan \beta \qquad (10.29)$$

When the rear wheels begin to climb a step, the relation between R, β and h is calculated (Fig. 10.10),

$$\cos \beta = \frac{R - h}{R}, \qquad (10.30)$$

where

$$\sin \beta = \sqrt{1 - \cos^2 \beta} \qquad (10.31)$$

From (10.30), (10.31), we obtain

$$\sin \beta = \frac{\sqrt{2Rh - h^2}}{R} \qquad (10.32)$$

From (10.29), (10.33), we obtain

$$\tan \beta = \frac{\sqrt{2Rh - h^2}}{R - h} \qquad (10.33)$$

Thus, from (10.29), (10.32),

$$\frac{f}{N_r} = \frac{\sqrt{2Rh - h^2}}{R - h} \qquad (10.34)$$

We define the proportional constant k_r $(0 \le k_r < 1)$,

$$h = k_r \cdot R \qquad (10.35)$$

(10.35) is substituted for (10.34) and we obtain (10.36),

$$\frac{f}{N_r} = \frac{\sqrt{k_r(2 - k_r)}}{1 - k_r} \qquad (10.36)$$

Using (10.36), we perform a numerical calculation to clarify the step climbing ability of the wheelchair rear wheels. The horizontal axis in Fig. 10.13 shows k_r and the vertical axis shows f / N_r. The shading in Fig. 10.13 shows $f / N_r < \mu$, which indicates the rear wheels are able to exert the driving force to the step. The step height at which the wheelchair rear wheels are able to climb is at most 25.7 % of the height of the rear wheel radius when the friction coefficient is $\mu = 0.9$, and is 10.6 % when the friction coefficient is $\mu = 0.5$.

10.2.4. Result of Simulations

Table 10.1 lists the results of the simulations. The table tells us that the wheelchair is not able to even climb a low step. For example, when the wheelchair moves in the area of $\mu = 0.5$ and encounters a step, it can climb a step height of 14.9 [mm] at most. It is not able to climb over a 27.5 [mm] height step because it would tip over backward even if the wheelchair moves in the area of $\mu = 0.9$ or 0.5.

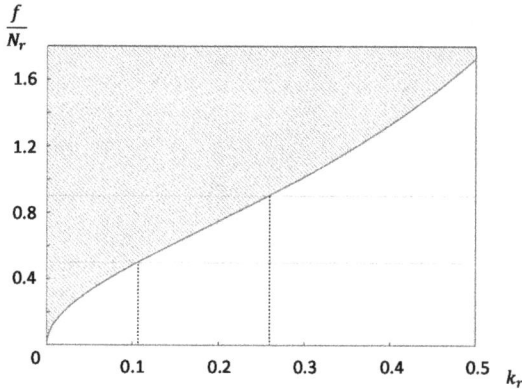

Fig. 10.13. Relation between step height and friction coefficient where rear wheels are able to climb.

In these simulations, as we described above, we assumed that the tires of the wheelchair do not transform, and the wheelchair user does not change his posture. If the user has the upper-body capability of an able-bodied person and is good at operating the wheelchair, it is actually possible to climb a higher step than the results of the simulation show. However, for most wheelchair users, it is clear that step climbing is a dangerous and difficult undertaking.

Table 10.1. Results of simulations of wheelchair step climbing.

Limitation of climbing	Step height [mm] (μ =0.5–0.9)
Front wheels	14.9–35.5
Tipping over backward	27.5
Rear wheels	31.7–77.0

10.3. Cooperative Step Climbing Using a Wheelchair and a Robot

As described above, wheelchair users face a lot of obstacles. Under most current circumstances, many wheelchair users need a caregiver. In any case, taking care of a wheelchair is hard work and we have to improve it. Our research group has studied the caregiving robot system, which helps a wheelchair user. In this section and the next, we begin by evaluating step climbing strategies for a wheelchair using a care robot.

10.3.1. Cooperative Step Climbing System

The robot used in this research is the wheeled "Tateyama" developed in this laboratory (Fig. 10.14). When climbing a step, the wheelchair and the robot are deployed in a forward-and-aft configuration (Fig. 10.15). Table 10.2 lists the specifications of the robot.

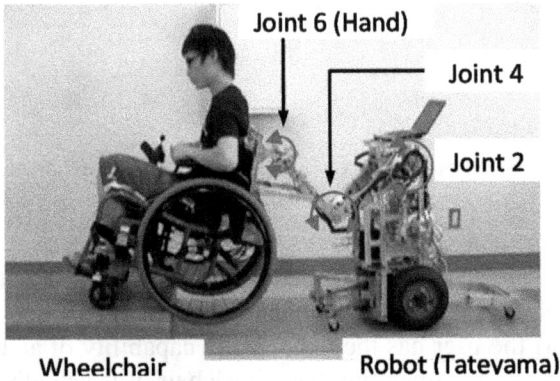

Fig. 10.14. Wheelchair and robot.

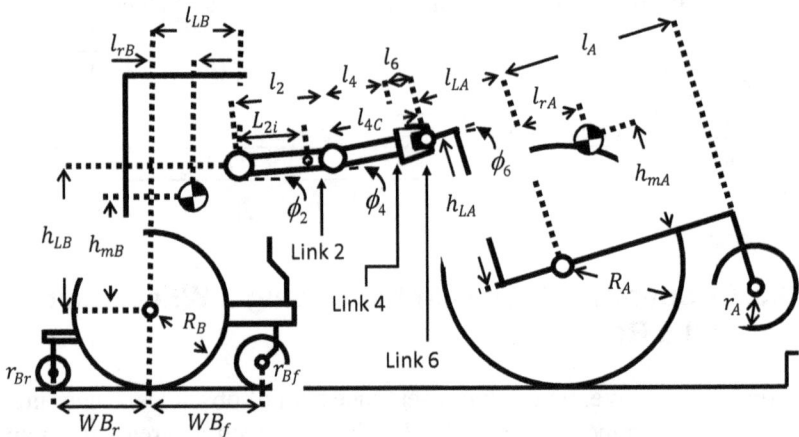

Fig. 10.15. Model of wheelchair during climbing of the front wheels.

Table 10.2. Robot specifications.

Overall length	230–800	[mm]
Overall height	747	[mm]
Radius of front wheels (r_{Bf})	25	[mm]
Radius of rear wheels (r_{Br})	19	[mm]
Wheelbase (W_{Bf})	190–440	[mm]
Wheelbase (W_{Br})	270	[mm]
Mass position from the rear axes (l_{rB})	93	[mm]
Height of the mass from the rear axes (h_{mB})	286	[mm]
Position of Joint 2 from the rear axes (l_{LB})	90	[mm]
Height of Joint 2 from the rear axes (h_{LB})	532	[mm]
Mass of the robot body	55	[kg]
Mass of link 2 (from Joints 2 to 4)	2.55 × 2	[kg]
Mass of link 4 (from Joint 4 to hand)	0.8 ×2	[kg]
Length of link 2 (l_2)	330	[mm]
Length of link 4 (l_4)	300	[mm]
Length of the hand (l_6)	105	[mm]
Length from Joint 4 to the connecting position (l_{4C})	370	[mm]
Mass position of link 2 (L_2)	67	[mm]
Mass position of link 4 (L_4)	169	[mm]
Mass position of link 6 (hand mechanism) (L_6)	35	[mm]

The robot has three pairs of wheels consisting of front, middle, and rear wheels on the left and right sides. The front and rear pairs are casters whose positions can be shifted and folded, and the middle pair are driving wheels (Fig. 10.16).

The robot has manipulators attached to the left and right sides of its upper half: each arm has 5 degrees of freedom (DOF) and each hand has 1 DOF for a total of 6 DOF (Fig. 10.17). In this study, the length from J_2 (shoulder joint: Joint 2) to J_4 (elbow joint: Joint 4) is called "Link 2" (length l_2), from J_4 to J_5 (wrist) is called "Link 4" (length l_4), and from J_5 (wrist) to the tip of the hand is called "Link 6" (length l_6). The length from the J_4 (elbow) to J_6 (location of the connection between the wheelchair and the robot: Joint 6) is designated l_{4C} (Fig. 10.15).

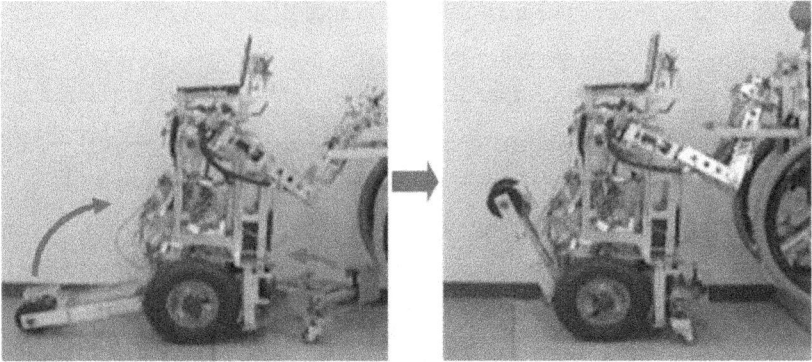

Fig. 10.16. Folding of front and rear wheels.

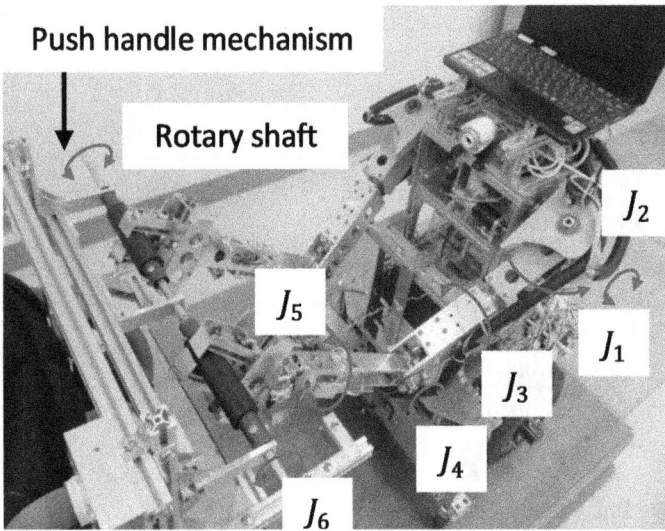

Fig. 10.17. Push handle mechanism of the wheelchair.

The manipulator joint angles are -90 [deg] $\leq \emptyset_2 \leq +90$ [deg] and 0 [deg] $\leq \emptyset_4 \leq +100$ [deg]. The hands consist of two fingers that open and close in order to hold objects (Fig. 10.17). These axes and the hands are driven by small DC motors (Joints 1, 2: 20 [W], Joints 3, 4: 6 [W], Joint 5: 2.5 [W], Joint 6 (hand): 1.5 [W]). The robot has a stopper mounted on its front as a part of its body (Fig. 10.18 (a)). The stopper enables the robot to imitate the operation whereby a human pushes an object by limiting the passive rotation about the shoulder joint as the upper arm is pushed into the chest (Fig. 10.18 (b)). As described below, this configuration limits the passive rotational travel of the manipulators when the robot has been pushed (Fig. 10.19 (a), (b)).

The wheelchair (NOVA Integral-ME) has a shape typical of wheelchairs available in the market (Fig. 10.14). Table 10.3 provides the specifications. This is a manually operated wheelchair to which an electric drive unit was added. In this study, the wheelchair is operated manually by the user, but also has a push handle mechanism on the back which is held by the robot hands (Fig. 10.17). The push handle mechanism is composed of a rotary shaft that allows passive rotation and a stopper for robot climbing. The hand mechanism grasps the shaft to connect the two vehicles. Angle $\emptyset_6$ is formed by the wheelchair with Link 4 (Fig. 10.15). The stopper of the wheelchair is composed of front and rear bars (Fig. 10.20 (a)), and is mounted on the rear side. During climbing of the robot, the sides of the robot are opened, and the two manipulators are inserted into the stopper (Fig. 10.20 (b)). Next, the robot pushes the front bars to lift the robot front wheels (Fig. 10.21 (a)). The rear bars are used to prevent the robot from tipping over backward when the robot inclines and its center of gravity shifts behind the contact point between the middle wheels and the ground (Fig. 10.21 (b)).

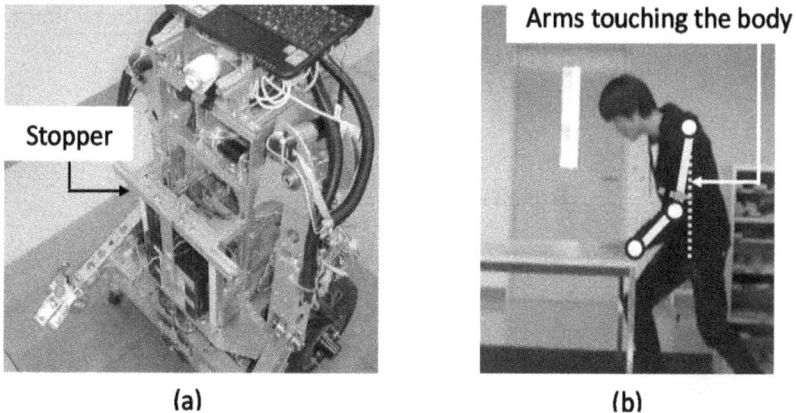

(a)   (b)

Fig. 10.18. Control of rotary motions of the robot shoulder by using the body: (a) The front body of the robot; (b) A human pushing an object.

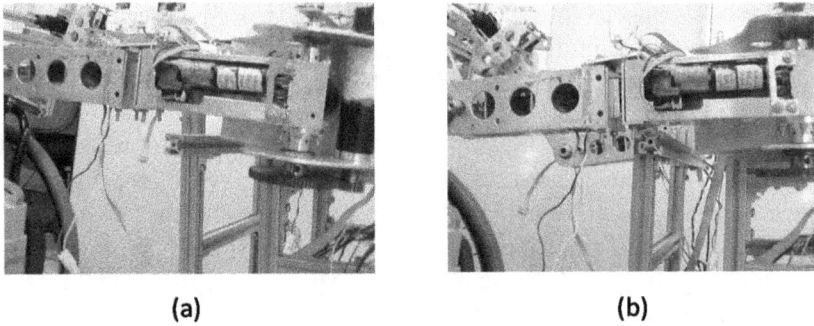

(a)   (b)

Fig. 10.19. Stopper of the robot: (a) Pulling the wheelchair;
(b) Pushing the wheelchair.

Table 10.3. Wheelchair specifications.

Overall length	1060	[mm]
Overall height	985	[mm]
Radius of front wheels (r_A)	63	[mm]
Radius of rear wheels (R_A)	300	[mm]
Wheelbase (l_A)	430	[mm]
Hand rim position (l_{LA})	250	[mm]
Mass position from the rear wheel axes (l_{rA})	149	[mm]
Height of mass from rear wheel axes (h_{mA})	371	[mm]
Mass (wheelchair + driver) (M_A)	92.7	[kg]

(a)   (b)

Fig. 10.20. Wheelchair stopper: (a) Climbing of the wheelchair;
(b) Climbing of the robot.

(a)   (b)

Fig. 10.21. Action of the wheelchair stopper: (a) Lifting the robot
front wheels; (b) Preventing the robot from falling down.

Fig. 10.22 is a diagram of the system configuration. The motors mounted
on the robot are connected to motion controllers (Faulhaber
MCDC3006-S, MCDC3003-S). In turn, these are connected to a
notebook PC mounted on the robot. The motors are controlled via
commands issued by the Faulhaber Motion Manager 4 software package.
The robot employs a camera built into the PC, and the moving images
from that camera and the Motion Manager 4 operating window are
displayed on the notebook PC mounted on the robot. The screen on this
notebook PC uses Real VNC software and is transmitted over the intranet
as-is to the display of the PC used by a caregiver at a different location.
The caregiver and the wheelchair user both wear headsets and use the
telecommunication software Skype to communicate verbally. The
caregiver's headset is connected to the caregiver's PC, and the wheelchair
user's headset is connected to the PC on the robot. The caregiver controls
the robot by operating Motion Manager 4 from his PC. The keyboard
commands for Motion Manager 4 are issued by JoyToKey software and
correspond to the manipulation by the caregiver for operating the robot.
The robot has internal and external sensors (encoders and touch sensors).
Thus, the robot is moved by the information integrated in the sensors'
signals with the commands from the caregiver.

10.3.2. Process of Moving Over a Step

When encountering a step, the robot hands grasp the rotary shaft of the
wheelchair push handle mechanism, thus linking the wheelchair and
robot. In this study, stage 1 and stage 2 signify the processes in which
the front and the back wheels of the wheelchair, respectively, ascend the
step. Similarly, stage 3 and stage 4 signify the processes in which the

front and the back wheels of the robot, respectively, ascend the step. Stage 1 is also divided into stages 1-1 and 1-2.

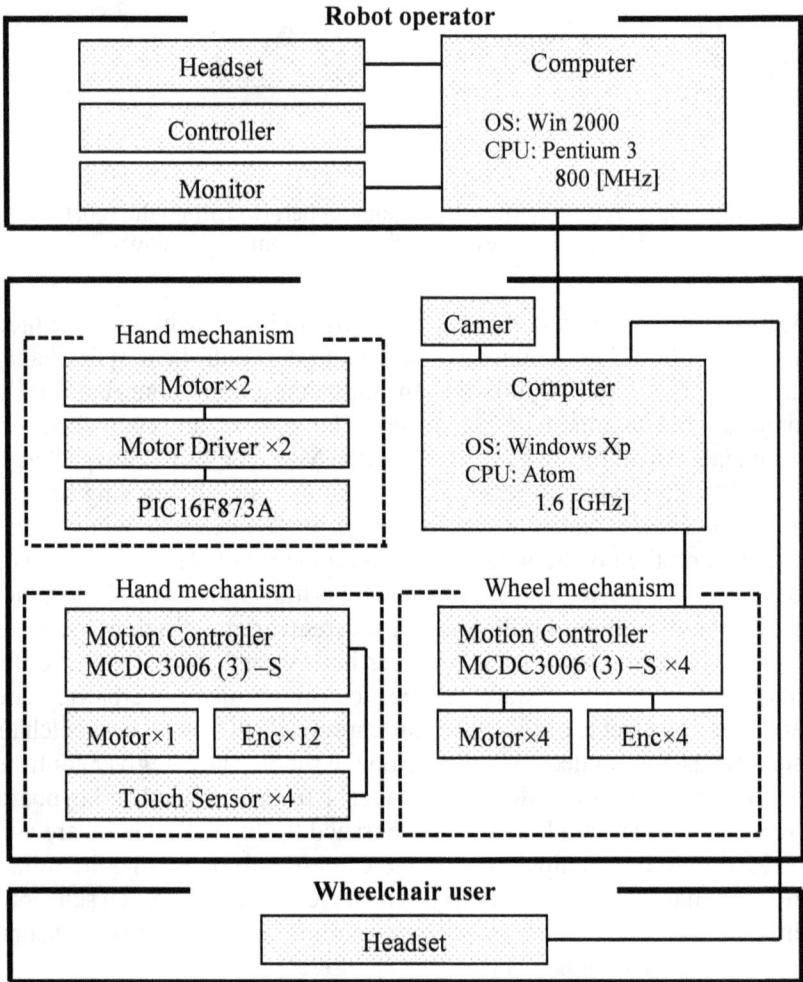

Fig. 10.22. Diagram of cooperative step climbing system.

The climbing processes are described below. The stages shown in Fig. 10.23 correspond to (1)–(16) below. "Forward" or "Backward" signifies the robot or the wheelchair motion ahead or behind, respectively, and "Free" is the state in which the vehicle does not do anything. "Stop" is the state in which the vehicle does not move.

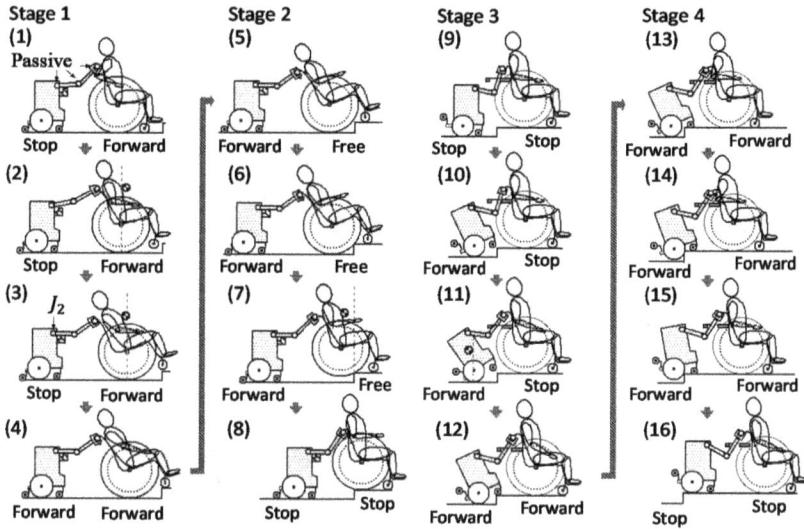

Fig. 10.23. Step climbing process of the wheelchair (stages 1–4).

[stage 1]

<stage 1-1>

(1) Joints 2, 4 and 6 are allowed to rotate passively until the ascent of the wheelchair has been completed. (2) The robot controller stops the robot. The wheelchair user manipulates the hand rims as if to move forward, and this action lifts the front wheels.

<stage 1-2>

(3) If the location of the center of gravity of the wheelchair shifts behind the contact point between the wheelchair rear wheels and the ground as the wheelchair tilt increases, the wheelchair exerts forces on the manipulators, and this causes passive rotation about Joint 2. In this case, the bottom part of the manipulator upper-arm link comes into contact with the stopper and limits the extent of the rotation (Fig. 10.19 (b)). Thus, the robot supports the wheelchair from behind to prevent the wheelchair from tipping over backward. (4) The robot moves forward and the wheelchair user manipulates the hand rims to adjust the difference between the speeds of the two vehicles, so that the front wheels of the wheelchair are placed on the upper step. After completion of stage 1, the wheelchair user does not perform any operations until the end of stage 2.

[stage 2]

(5) The robot continues to move forward while pushing the wheelchair from behind. (6) The back wheels of the wheelchair come into contact with the step. (7) The robot continues to push on the wheelchair so that the rear wheels of the wheelchair climb up onto the step. The robot supports the wheelchair during this process to prevent the wheelchair from tipping over backward. (8) Once the wheelchair rear wheels have reached the upper step, the robot stops.

[stage 3]

(9) After stage 2, the rear wheels of the robot are folded upward (Fig. 10.16). The sides of the robot are opened, and the two manipulators are inserted into the stopper (Figs. 10.20 (a) and (b)). The wheelchair user holds the hand rims and maintains the position of the wheelchair. The robot moves forward, and the manipulator forearm link comes into contact with the stopper of the wheelchair (Fig. 10.21 (a)). (10) The robot continues to push on the wheelchair and the front wheels of the robot are lifted. (11) If the location of the center of gravity of the robot shifts behind the contact point between the middle wheels of the robot and the ground as the robot tilt increases, the robot begins to tip over backward, but part of the manipulator forearm link comes into contact with the stopper of the wheelchair and limits the extent of rotation (Fig. 10.21 (b)). Thus, the wheelchair supports the robot and prevents the robot from tipping over backward. (12) The robot moves forward and the wheelchair user manipulates the hand rims, and thereby adjusts the difference between the speeds of the two vehicles, so that the front wheels of the robot are placed on the upper step.

[stage 4]

(13) Both vehicles move forward. (14) The middle wheels of the robot come into contact with the step. The wheelchair pulls the robot, and the value of the normal reaction from the step on the robot middle wheels (driving wheels) is increased. Consequently, the robot is able to avoid falling down by the force of the manipulators. The middle wheels of the robot start to climb the step. (15) Both vehicles continue to move forward. (16) The middle wheels of the robot are able to climb the step. Once the robot middle wheels have reached the upper step, both vehicles are stopped.

10.4. Theoretical Analysis of Cooperative Step Climbing Using the Wheelchair and the Robot

The purpose of this section is to illustrate the theoretical analysis method by using stage 1 and stage 2 as examples.

That is, this method is analyzed by considering the statics of the wheelchair and the robot climbing a step at slow speed and keeping their balance. We assumed that both vehicles can generate enough power to climb a step.

Fig. 10.24 shows the model of the vehicles when the wheelchair center of gravity is in front of the contact position between the rear wheels and the ground. Fig. 10.25 shows the model when the wheelchair center of gravity is behind the contact position. Σ_B is the basic coordinate system for the robot; contact point B is between the robot middle (driving) wheels and the origin is the ground (Fig. 10.26). Joints 2 (shoulder), 4 (elbow), and 6 (location where the hands hold the push handle) are controlled passively. The position vectors for these joints in system Σ_B are expressed as ${}^B\boldsymbol{p}_{2i} = [x_{2i} \ z_{2i}]^T$ (i = 1–3). Table 10.4 lists the position vectors in the system. In the same way, the position vectors for the contact points between the robot front and rear wheels and the ground are expressed as ${}^B\boldsymbol{p}_{fwb}$ and ${}^B\boldsymbol{p}_{rwb}$, respectively. The body of the robot, if we neglect the manipulators, is Link 0 with mass m_0. If the centers of mass of the robot body and each manipulator link (Links 2, 4, and 6) are denoted by ${}^B\boldsymbol{p}_{gj}$ (where j = 0–3), then the center of gravity of the entire robot is ${}^B\boldsymbol{p}_{GB}$.

Table 10.5 lists the force vectors in the system. The driving force vector for the robot middle wheels is $\boldsymbol{f}_1$, and the resistance force from the ground surface is $\boldsymbol{f}_2$. Additionally, the resistance force at the robot front wheels is $\boldsymbol{f}_3$, and that at the rear wheels is $\boldsymbol{f}_4$. The reaction force from the linked wheelchair is given by $\boldsymbol{f}_L$.

Σ_A for the wheelchair is the coordinate system fixed at the point of contact between the wheelchair rear wheels and the ground, point A (Fig. 10.26). In Σ_A, the wheelchair center of gravity is located at ${}^A\boldsymbol{p}_{GA}$, and the push handle location (where the wheelchair is held by the robot hand). The push handle position P_C is ${}^A\boldsymbol{p}_C$.

The driving force at the wheelchair rear wheels is f_5, and the resistance force felt at the ground surface is f_6. Furthermore, the reaction force from the linked robot is given by f_L'.

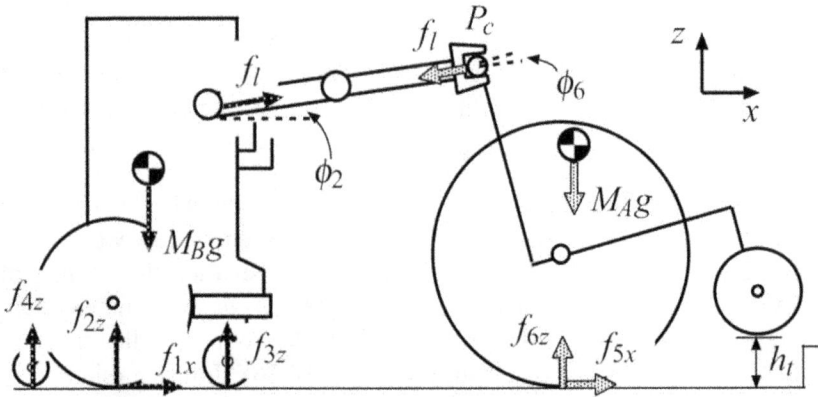

Fig. 10.24. Cooperative step climbing system (Stage 1, the wheelchair center of gravity is in front of the contact position between the rear wheels and the ground).

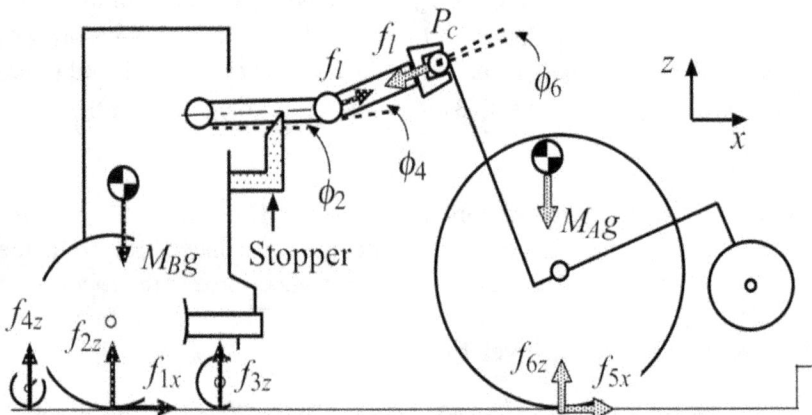

Fig. 10.25. Cooperative step climbing system (Stage 1, the wheelchair center of gravity is behind the contact position between the rear wheels and the ground).

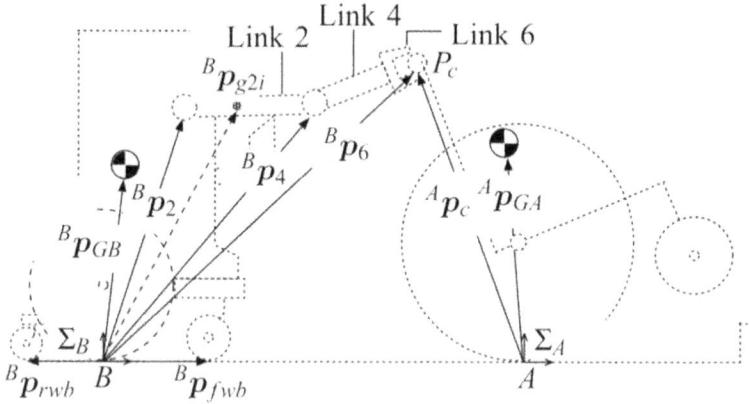

Fig. 10.26. Position vectors for the joints in the system.

Table 10.4. Position vectors in the system.

Position vector of shoulder (Joint 2) in Σ_B	$^B\mathbf{p}_2 = [x_2 \quad z_2]^T = [l_{LB} \quad R_B + h_{LB}]^T$
Position vector of elbow (Joint 4) in Σ_B	$^B\mathbf{p}_4 = [x_4 \quad z_4]^T$ $= \begin{bmatrix} l_{LB} + l_2 \cos\varphi_2 \\ R_B + h_{LB} + l_2 \sin\varphi_2 \end{bmatrix}$
Position vector of twist (Joint 6) in Σ_B	$^B\mathbf{p}_6 = [x_6 \quad z_6]^T$ $= \begin{bmatrix} l_{LB} + l_2 \cos\varphi_2 + l_{4C} \cos(\varphi_2 + \varphi_4) \\ R_B + h_{LB} + l_2 \sin\varphi_2 + l_{4C} \sin(\varphi_2 + \varphi_4) \end{bmatrix}$
Position vectors for the contact points between the robot front wheels and the ground	$^B\mathbf{p}_{fwb} = [WB_f \quad 0]^T$
Position vectors for the contact points between the robot rear wheels and the ground	$^B\mathbf{p}_{rwb} = [-WB_r \quad 0]^T$
Centers of mass of the robot body and each manipulator link (Links 2, 4, and 6)	$^B\mathbf{p}_{g\,j} = [x_{g2j} \quad z_{g2j}]^T \quad (j = 0-3)$
Center of gravity of the entire robot	$^B\mathbf{p}_{GB} = [x_{GB} \quad z_{GB}]^T = \dfrac{\sum_{j=0}^{3} m_{2j}^B \mathbf{p}_{g2j}}{\sum_{j=0}^{3} m_{2j}^B}$

275

Table 10.4. (Continued). Position vectors in the system.

Wheelchair center of gravity	$^A\mathbf{p}_{GA} = [x_{GA} \quad z_{GA}]^T$ $= \begin{bmatrix} l_{rA}\cos(\varphi_2 + \varphi_4 + \varphi_6) - h_{mA}\sin(\varphi_2 + \varphi_4 + \varphi_6) \\ l_{rA}\sin(\varphi_2 + \varphi_4 + \varphi_6) + h_{mA}\cos(\varphi_2 + \varphi_4 + \varphi_6) + R_A \end{bmatrix}$
Push handle position (P_C)	$^A\mathbf{p}_C = [x_C \quad z_C]^T$ $= \begin{bmatrix} -l_{LA}\cos(\varphi_2 + \varphi_4 + \varphi_6) - h_{LA}\sin(\varphi_2 + \varphi_4 + \varphi_6) \\ -l_{LA}\sin(\varphi_2 + \varphi_4 + \varphi_6) + h_{LA}\cos(\varphi_2 + \varphi_4 + \varphi_6) + R_A \end{bmatrix}$
Contact position between the wheelchair rear wheels and the ground (A) in Σ_B	$^B\mathbf{p}_A = [d \quad 0]^T$

Table 10.5. Force vectors in the system.

Driving force vector for the robot middle wheels	$\mathbf{f}_1 = [f_{1x} \quad 0]^T$
Resistance force from the ground surface	$\mathbf{f}_2 = [0 \quad f_{2z}]^T$
Resistance force at the robot front wheels	$\mathbf{f}_3 = [0 \quad f_{3z}]^T$
Resistance force at the robot rear wheels	$\mathbf{f}_4 = [0 \quad f_{4z}]^T$
Reaction force from the linked wheelchair	$\mathbf{f}_L = [f_l \cos(\varphi_2 + \varphi_4) \quad f_l \sin(\varphi_2 + \varphi_4)]^T$
Driving force at the wheelchair rear wheels	$\mathbf{f}_5 = [f_{5x} \quad 0]^T$
Resistance force at the wheelchair rear wheels felt at the ground surface	$\mathbf{f}_6 = [0 \quad f_{6z}]^T$
Reaction force from the linked robot	$\mathbf{f}_L' = [-f_l \cos(\varphi_2 + \varphi_4) \quad -f_l \sin(\varphi_2 + \varphi_4)]^T$
Gravitational acceleration vector	$\boldsymbol{g} = [0 \quad -g]^T$

10.4.1. Requirement of the Manipulator Angles to Avoid Collision and to Grasp the Push Handle

The achievement of this cooperative step climbing method is much influenced by the angle of the manipulator links. In this chapter, theoretical analyses clarify that the manipulator angles are able to avoid collision with the link parts of the manipulator (Fig. 10.27, A1), and are able to both grasp using the robot hands (Fig. 10.28, A2 and A3) and to preventing slippage of driving wheels (Fig. 10.29, B1-B4).

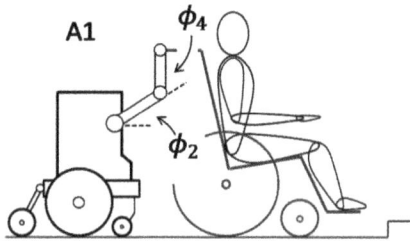

Fig. 10.27. Avoidance of collision between the wheelchair and the links of the manipulators.

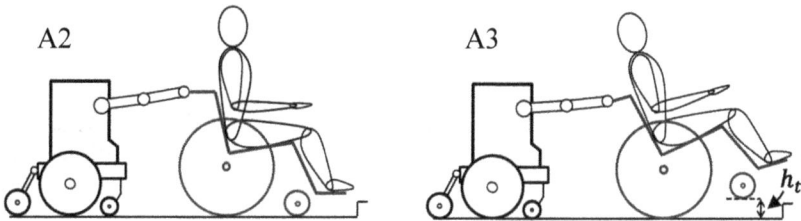

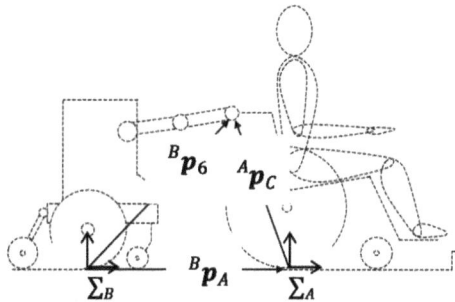

Fig. 10.28. Push handle position that the hand mechanism is able to grasp.

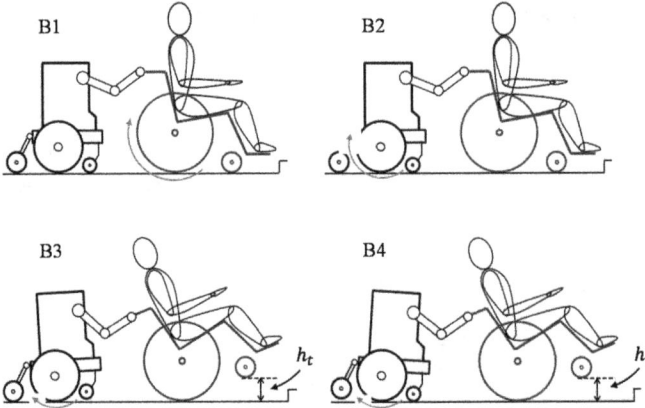

Fig. 10.29. The ϕ_2 and ϕ_4 that exert the driving force on the ground (stage 1).

[Requirement: A1]

The system has to avoid a collision between the wheelchair and the links of the manipulator (Fig. 10.27, A1). The requirement is described below.

$$\phi_2 + \phi_4 < 90 \text{ [deg]} \tag{10.37}$$

[Requirement: A2]

The robot hands have to grasp the wheelchair push handle when the wheelchair front wheels begin to lift up (wheelchair inclination is 0, Fig. 10.28, A2).

$$\left\| {}^B\mathbf{p}_6 \right\| \geq \left\| {}^B\mathbf{p}_A + {}^A\mathbf{p}_C \right\|, \, (\phi_2 + \phi_4 + \phi_6 = 0) \tag{10.38}$$

[Requirement: A3]

The robot hands have to grasp the wheelchair push handle when the wheelchair inclination is maximum (wheelchair inclination is 24.54 [deg], Fig. 10.28, A3).

Here, 24.54 [deg] is the maximum incline of the wheelchair when the wheelchair operator climbs a step, and the height between the lowest point on the front wheel tread surface and the ground surface below the step is $h_t = 0.2$ [m] (Fig. 10.28). It was observed that people tend to raise the front wheels higher than the step they intend to traverse when actually operating a wheelchair. Thus, h_t was measured for five participants and the results were used when specifying a maximum tilt angle.

$$\left\| ^B \mathbf{p}_6 \right\| \geq \left\| ^B \mathbf{p}_A + ^A \mathbf{p}_C \right\|, (\phi_2 + \phi_4 + \phi_6 = 24.54 \, [\text{deg}]) \quad (10.39)$$

10.4.2. Requirement to Exert Enough Driving Force on the Ground to Climb a Step

The requirements of preventing slippage in stage 1 are listed below (Fig. 10.29, B1–B4), The forces $f_{1x}, f_{2z}, \cdots, f_{6z}$ are shown in Figs. 10.24 and 10.25. Fig. 10.24 shows the stage 1 state in which the wheelchair center of gravity is forward of the contact point between the rear wheels and the ground (Fig. 10.23 (1)–(2)). At this time point, the robot is stopped, the wheelchair is propelled, and the robot exerts a backward force by pulling on the wheelchair. Fig. 10.25 shows the state when the tilt of the wheelchair is increasing, and the wheelchair center of gravity is behind the contact point between the rear wheel and the ground (Fig. 10.23 (3)–(4)). In stage 1, the situation shown in Fig. 10.24 changes to the situation shown in Fig. 10.25. In this procedure, the stoppers of the robot limit the amount of passive rotation about the robot shoulder joint (Figs. 10.19 (a) and (b)).

We note that the robot has not only middle wheels (driving wheels) but also rear wheels (casters). Thus, we do not have to analyze the situation of tipping over backward because slippage of the middle wheels (driving wheels) would occur before tipping over backward in this system.

[Requirement: B1]

In stage 1, the wheelchair rear wheels (driving wheels) do not slip, and they exert enough driving force on the ground to lift the wheelchair front wheels.

$$\mu > \frac{|f_{5x}|}{f_{6z}} \quad (\phi_2 + \phi_4 + \phi_6 = 0) \quad (10.40)$$

[Requirement: B2]

In stage 1, the robot middle wheels (driving wheels) do not slip, and they exert enough driving force on the ground to lift the wheelchair front wheels.

$$\mu > \frac{|f_{1x}|}{f_{2z}} \quad (\phi_2 + \phi_4 + \phi_6 = 0) \qquad (10.41)$$

[Requirement: B3]

In stage 1, the robot middle wheels do not slip when the wheelchair inclines and the robot supports the wheelchair (Figs. 10.25 and 10.29 B3). The robot is pushed backward by the tilting wheelchair; however, the robot is able to remain stopped if the robot front wheels are lifted, $f_{3z} = 0$.

$$\mu > \frac{|f_{1x}|}{f_{2z}} \quad (\phi_2 + \phi_4 + \phi_6 = 24.54 \text{ [deg]}) \qquad (10.42)$$

[Requirement: B4]

In stage 1, the robot middle wheels do not slip when the wheelchair inclines and the robot supports the wheelchair (Figs. 10.25 and 10.29 B4). The robot is pushed backward by the tilting wheelchair; however, the robot is able to remain stopped if the robot rear wheels are lifted, $f_{4z} = 0$.

$$\mu > \frac{|f_{1x}|}{f_{2z}} \quad (\phi_2 + \phi_4 + \phi_6 = 24.54 \text{ [deg]}) \qquad (10.43)$$

Similarly, the requirements of preventing slippage in stage 2 are listed below (Fig. 10.30, B5, B6).

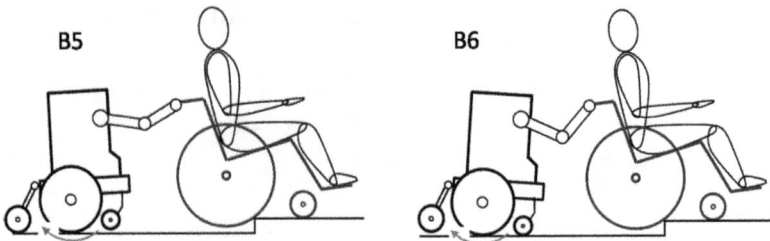

Fig. 10.30. The ϕ_2 and ϕ_4 that exert the driving force on the ground (stage 2).

Here, 15.08 [deg] is the incline of the wheelchair when the wheelchair rear wheels begin to climb the step.

[Requirement: B5]

In stage 2, the robot middle wheels do not slip when the wheelchair rear wheels climb a step. The robot is pushed backward by the wheelchair ($f_{3z} = 0$); however, the robot is able to move forward if the robot front wheels are lifted.

$$\mu > \frac{|f_{1x}|}{f_{2z}} \quad (\phi_2 + \phi_4 + \phi_6 = 15.08 \text{ [deg]}) \qquad (10.44)$$

[Requirement: B6]

In stage 2, the robot middle wheels do not slip when the wheelchair rear wheels climb a step. The robot is pushed backward by the wheelchair ($f_{4z} = 0$); however, the robot is able to move forward if the robot rear wheels are lifted.

$$\mu > \frac{|f_{1x}|}{f_{2z}} \quad (\phi_2 + \phi_4 + \phi_6 = 15.08 \text{ [deg]}) \qquad (10.45)$$

10.4.3. Theoretical Analysis of Cooperative Step Climbing Using the Wheelchair and the Robot

Summing the total forces on the wheelchair exerted by the ground surface (resistance) and by the linked robot for $f_{\Sigma A} \in R^2$ (Figs. 10.24 and 10.25), we find that

$$f_{\Sigma A} = \left[f_{5x} - f_l \cos\left(\phi_2 + \phi_4\right) \quad f_{6z} - f_l \sin\left(\phi_2 + \phi_4\right) \right]^T (10.46)$$

When the linked vehicles are moving together in static equilibrium, the equilibrium for both the x and z axes yields (10.47), while the equilibrium of moments about the point of contact between the wheelchair rear wheels and the ground yields (10.48). Here, g is the gravitational acceleration (Table 10.5).

$$f_{\Sigma A} + M_A g = 0 \qquad (10.47)$$

$$^{A}\boldsymbol{p}_{GA} \times M_{A}\boldsymbol{g} + {}^{A}\boldsymbol{p}_{C} \times \boldsymbol{f}_{L}{}' = 0 \qquad (10.48)$$

We obtain (10.49) and (10.50) from (10.47),

$$f_{5x} = f_{l}\cos\left(\phi_{2} + \phi_{4}\right) \qquad (10.49)$$

$$f_{6z} = M_{A}g + f_{l}\sin\left(\phi_{2} + \phi_{4}\right) \qquad (10.50)$$

Then, from (10.48), we find

$$f_{l} = \frac{x_{GA}M_{A}g}{z_{c}\cos\left(\phi_{2} + \phi_{4}\right) - x_{c}\sin\left(\phi_{2} + \phi_{4}\right)} \qquad (10.51)$$

Next, from the z-coordinate of $^{B}p_{6}$ and $^{A}p_{C}$, we obtain

$$h_{LA} = \frac{R_{B} + h_{LB} + l_{2}\sin\phi_{2} + l_{4C}\sin\left(\phi_{2} + \phi_{4}\right) + l_{LA}\sin\left(\phi_{2} + \phi_{4} + \phi_{6}\right) - R_{A}}{\cos\left(\phi_{2} + \phi_{4} + \phi_{6}\right)} \qquad (10.52)$$

When the robot acts statically in stage 1, (10.53) holds, and the equilibrium in the x and z axes gives us

$$\boldsymbol{f}_{\Sigma B} + M_{B}\boldsymbol{g} = 0, \qquad (10.53)$$

where $\boldsymbol{f}_{\Sigma B} \in R^{2}$ is the sum of the forces on the robot due to resistance at the ground surface and from the linked wheelchair, and

$$\boldsymbol{f}_{\Sigma B} = [f_{1x} + f_{l}\cos\left(\phi_{2} + \phi_{4}\right) \quad \sum_{k=2}^{4} f_{kz} + f_{l}\sin\left(\phi_{2} + \phi_{4}\right)]^{T} \qquad (10.54)$$

From (10.53), we obtain (10.55) and (10.56),

$$f_{1x} = -f_{l}\cos\left(\phi_{2} + \phi_{4}\right) \qquad (10.55)$$

$$f_{2z} = M_{B}g - f_{l}\sin\left(\phi_{2} + \phi_{4}\right) - f_{3z} - f_{4z} \qquad (10.56)$$

During the process of moving over a step, while the manipulators are pulling the robot wheelchair (Fig. 10.24), the manipulators and stoppers do not come into contact (Fig. 10.19 (a)). By the equilibrium of moments about the contact point between the robot driving wheel and the ground during this time, we obtain

$$^B p_{GB} \times M_B g + {}^B p_2 \times f_L + {}^B p_{fwb} \times f_3 + {}^B p_{rwb} \times f_4 = 0 \quad (10.57)$$

From (10.57), (10.58) is obtained,

$$x_{GB} M_B g - f_l \{ x_2 \sin(\phi_2 + \phi_4) - z_2 \cos(\phi_2 + \phi_4) \} + W B_f f_{4z} - W B_f f_{3z} = 0 \quad (10.58)$$

When the robot is supporting the wheelchair from behind (Fig. 10.25), passive rotation about Joint 2 (shoulder) is limited by the stopper (Fig. 10.19 (b)). At such time, Link 2 (the upper arm of the manipulator) can be treated as a part of the robot body and the equilibrium of moments about the contact point between the robot driving wheels is represented as

$$^B p_{GB} \times M_B g + {}^B p_4 \times f_L + {}^B p_{fwb} \times f_3 + {}^B p_{rwb} \times f_4 = 0 \quad (10.59)$$

From (10.59), (10.60) is obtained,

$$x_{GB} M_B g - f_l \{ x_4 \sin(\phi_2 + \phi_4) - z_4 \cos(\phi_2 + \phi_4) \} + W B_f f_{4z} - W B_f f_{3z} = 0 \quad (10.60)$$

We obtain f_{3z} and f_{4z} from (10.58) and (60). In addition, $\mu > |f_{1x}|/f_{2z}$ or $\mu > |f_{5x}|/f_{6z}$ can be calculated from (10.49)–(10.52), (10.55) and (10.56).

10.4.4. Simulation

The requirements in stages 1 and 2 were calculated above. The requirements except B3 and B4 limit the manipulator link angle in stages 1 and 2 (Fig. 10.31), where the coefficient of friction is $\mu = 0.72$ in the simulation. In this figure, the horizontal axis shows ϕ_2 and the vertical axis shows ϕ_4. The united shading S1 and S2 show the combination of link angles that are able to achieve stage 1. The shading S2 shows the combination of the link angles that are able to achieve stage 2.

The system is able to choose the link angle from the result of this simulation to achieve stage 1-1, stage 1-2 and stage 2 (see Section 10.5).

The next section presents the experiment of cooperative step climbing using the wheelchair and the robot.

Although they are not discussed here, stages 3 and 4 (the robot climbing process) are able to be calculated for each requirement by a similar method.

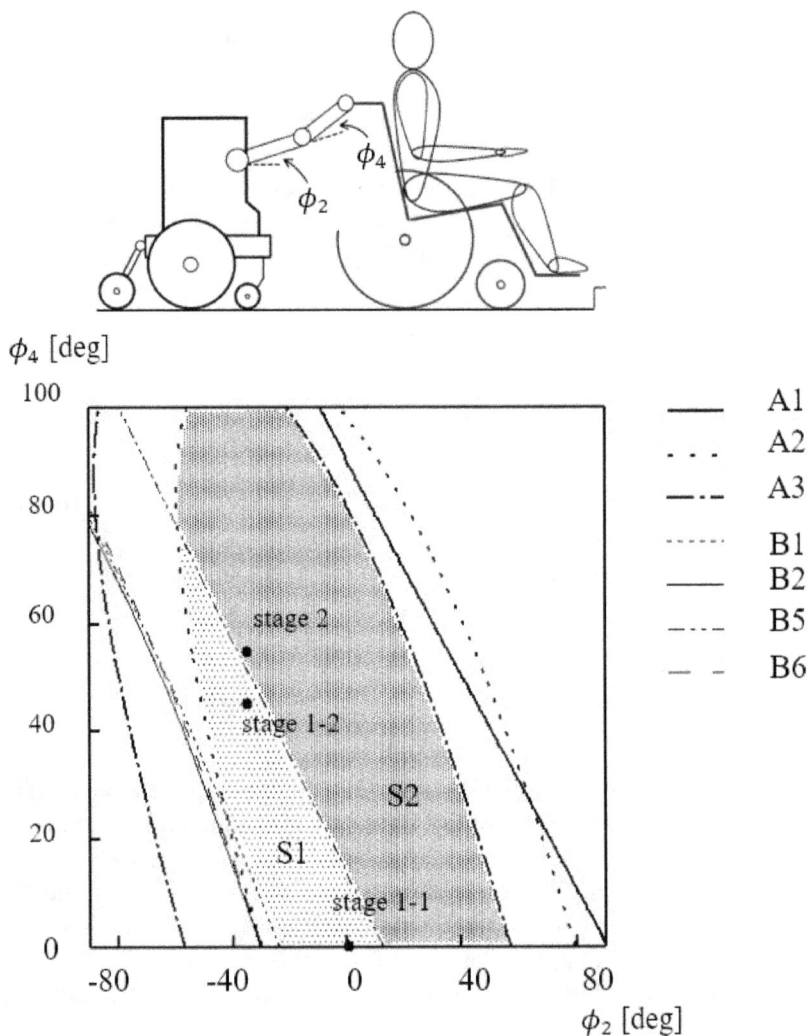

Fig. 10.31. Combinations of ϕ_2 and ϕ_4 by which the system achieves stages 1 and 2 ($\mu = 0.72$).

10.5. Experiment of Cooperative Step Climbing

An experiment was carried out by using this system under an environment of step height 120 [mm] and friction coefficient $\mu = 0.72$. The link angles ϕ_2 and ϕ_4 were set by using the results of the simulation (Fig. 10.32). The wheelchair user and robot operator were both able-bodied adult males. The wheelchair user and the robot were placed on one floor of the National Institute of Technology, Toyama College, and the robot operator was on another floor of the same building. The robot operator performed his task over the intranet while observing the situation via a camera and communicating with the wheelchair user over a voice link.

If the wheelchair was too close to the step in stage 1, the front wheels bumped into the vertical riser of the step. However, following instructions from the wheelchair user, the robot controller was able to back the two linked vehicles together and re-start the ascent. It was then possible for the front wheels of the wheelchair to climb the step with ease.

Subsequently, during stage 2, the user never needed to push the wheels. Specifically, it was possible to lift the chair onto the upper step by following the procedure proposed above and using only the forward operation of the robot. Stages 3 and 4 were then executed.

In the climbing processes, both vehicles inclined in turn. However, because this system only has one camera installed on the robot, the visual information provided was limited and the robot operator experienced some difficulty in controlling the robot. Thus, it is clear that the construction of a system to support the robot operator and the wheelchair operator based on exterior and perhaps other sensors is required in the future.

10.6. Conclusions

This chapter described a step climbing strategy and a theoretical analysis method of step climbing for a typical wheelchair and indicated the difficulty of step climbing.

We showed the cooperative step climbing strategy of the wheelchair and the robot, the robot and wheelchair system, and the theoretical analysis.

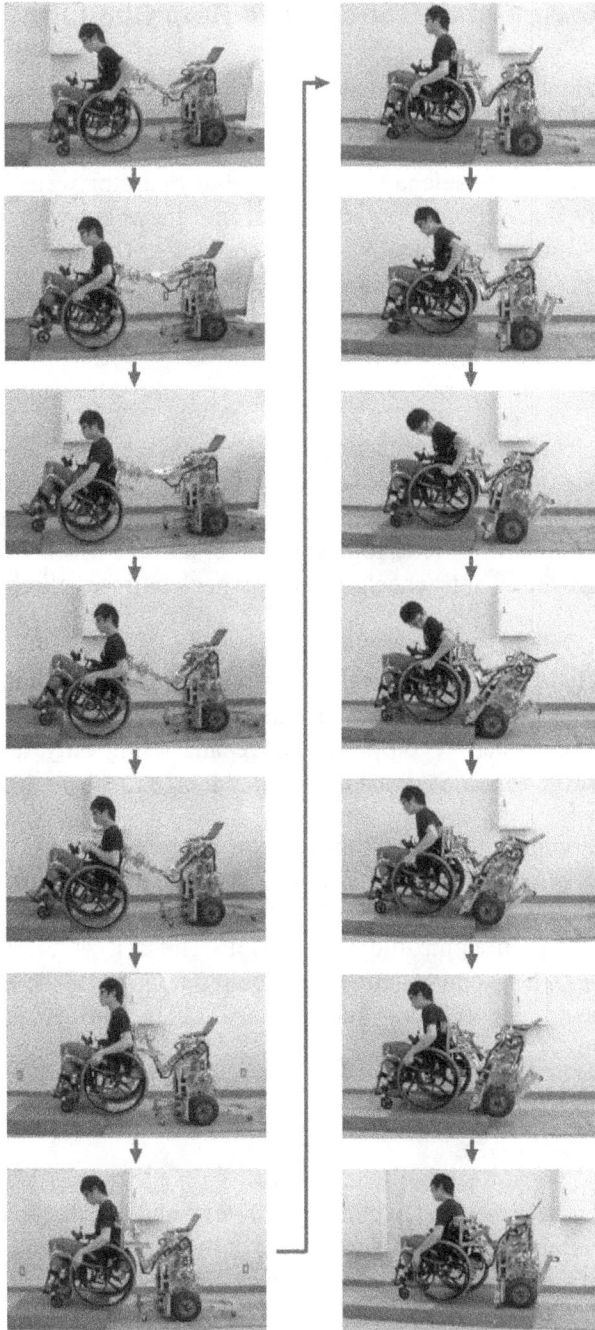

Fig. 10.32. Experiment of cooperative step climbing.

An experiment was carried out that incorporated teleoperation of the robot over an intranet. The effectiveness of the handling method for a heavy object was demonstrated by using a robot with manipulators driven by small motors. During the climbing processes, these vehicles inclined their front wheel and then climbed the step. However, because this system has only one camera installed on the robot, the visual information supplied to the robot operator was limited. Thus, it is clear that the robot operator will need an enhanced support system that can indicate the distance from the step and show other situations related to the vehicle. It is also clear that construction of a system to support the wheelchair operator based on exterior and perhaps other sensors is required in the future.

Despite the above observations, it is worthwhile to demonstrate that mobile manipulators, which are driven by small motors, are capable of handling a heavy cart (wheelchair) by pressing the manipulator links against the vehicle in addition to the grasping the push handle.

In the future, we will verify the force necessary to operate the wheelchair by this method and we will build an autonomous control system to assist the wheelchair user and robot operator.

Acknowledgments

This work was supported in part by a grant from the Daiwa Securities Health Foundation (2007, Grant No. 19), the Okawa Foundation for Information and Telecommunications (2008, Grant No. 08-18).

My thanks are also due to my parents Yoshihiro and Saiko, and my wife Erika.

References

[1]. Rory A. Cooper, Wheelchair Selection and Configuration, Medical Publishing (Japanese language edition by Igaku-Shoin, Ltd.), 1998.
[2]. C. Jill Calder, R. Lee Kirby, Fatal Wheelchair- related Accidents in the United State, *American Journal of Physical Medicine & Rehabilitation*, Vol. 69, No. 4, 1990, pp. 184-190.
[3]. V. Kumar, V. Krovi, Optimal traction control in a wheelchair with legs and wheels, in *Proceedings of the 4th National Applied Mechanics and Robotics Conference*, December 1995, pp. 95-030-01–95-030-07.

[4]. N. Yanagihara, F. Sugasawa, N. Suzuki, T. Ikeda, Y. Kanaumi, Mechanical analysis of a stair-climbing wheelchair using rotary cross arm with wheels, in *Proceedings of the 17th Annual Conference of the Robotics Society of Japan*, September 1999, pp. 1143-1144.

[5]. G. Quaglia, W. Franco, R. Oderio, Wheelchair. q, a motorized wheelchair with stair climbing ability, *Mechanism and Machine Theory,* Vol. 46, Issue 11, 2011, pp. 1601-1609.

[6]. M. Lawn, T. Ishimatsu, Modeling of a stair-climbing wheelchair mechanism with high single step capability, *IEEE Transactions on Neural Systems and Engineering*, Vol. 11, No. 3, 2003, pp. 323-332.

[7]. Y. Takahashi, S. Ogawa, S. Machida, Human assist robot (1st report: Prototype of wheelchair which can fly up and run), in *Proceedings of the JSME ROBOMEC' 99*, Tokyo, 1999, 1A1-75-106.

[8]. Independence Technology, L.L.C., iBOT (Online, retrieved: February 2013, http://www.ibotnow.com/, 2008.)

[9]. K. Taguchi, H. Sato, A study of the wheel-feet mechanism for stair climbing, *Journal of Robotics Society of Japan*, Vol. 15, 1997, pp. 118-123. (in Japanese).

[10]. K. Sugiyama, T. Ishimatsu, T. Shigechi, M. Kurihara, Development of stair-climbing machines at Nagasaki, in *Proceedings of 3rd International Workshop on Advanced Mechatronics,* Kanwon, Korea, 1999, pp. 214-217.

[11]. Y. Mori, K. Katsumura, K. Nagase, A pair of step-climbing units for a manual wheelchair user: Passing over several steps using a pair of portable slopes, *Advances in Mechanical Engineering*, Vol. 9, No. 3, 2017, pp. 1-11.

[12]. H. Ikeda, Y. Katsumata, M. Shoji, T. Takahashi, E. Nakano, Cooperative strategy for a wheelchair and a robot to climb and descend a step, *Advanced Robotics*, Vol. 22, 2008, pp. 1439-1460.

[13]. H. Ikeda, H. Kanda, N. Yamashima, E. Nakano, Step climbing and Descending for a Manual wheelchair with a Network care robot, in *Proceeding of the 2nd International Conference on Intelligent System and Applications (INTELLI'13)*, Venice, Italy, 21-26 April, 2013, pp. 95-102.

[14]. H. Asama, M. Sato, N. Goto, H. Kaetsu, A. Matsumoto, I. Endo, Mutual Transportation of Cooperative Mobile Robots, Using Forklift Mechanisms, in *Proceedings of the IEEE International Conference on Robotics and Automation (ICRA'96)*, Minneapolis, USA, 22-28 April 1996, pp.1754-1759.

[15]. H. Ikeda, N. Yamanaka, T. Kurose, S. Nagai, H. Doba, S. Kasuga, K. Sato, E. Nakano, Step Climbing of a Wheelchair Using a Wheeled Robot with Passive Joint Manipulators, *Journal of the Robotics Society of Japan*, Vol. 28, No. 7, 2010, pp. 802-810. (in Japanese).

Index